Propagation of Horticultural Crops

Propagation of Horticultural Crops

R.C. Upadhyaya

ANMOL PUBLICATIONS PVT. LTD.
NEW DELHI - 110 002 (INDIA)

ANMOL PUBLICATIONS PVT. LTD.
H.O.: 4374/4B, Ansari Road, Darya Ganj,
New Delhi-110 002 (India)
Ph.: 23278000, 23261597
B.O.: No. 1015, Ist Main Road, BSK IIIrd Stage
IIIrd Phase, IIIrd Block,
Bangalore - 560 085 (India)
Visit us at: www.anmolpublications.com

Propagation of Horticultural Crops

First Published, 2008

ISBN 978-81-261-3426-7

PRINTED IN INDIA

Printed at Mehra Offset Press, Delhi.

Contents

Preface

Horticulture has a special significance for the economic development. Most of the people, though hard working and face the calamities of nature and live in harsh conditions, yet produce lot of fruits, vegetables and other flowering and ornamental plants including the medicinal and aromatic plants. In order to increase the productivity of horticultural crops growers must take the advantage of the expertise of scientists. Keeping this idea in focus, the present work has been compiled which contains experiences of the experts and related people. The volumes cover advances in strategies, production, plant protection, value addition and other important areas of horticulture development.

The Encyclopaedia provides practical information on production technologies of tropical and subtropical fruits as well as plantation crops and horticultural nursery. Post-harvest management, marketing and export trends have been furnished for guidance. It also presents most realistic cost estimates and returns from cultivation of such crops. The volumes aim at viewing commercial prospects for horticulture. It has been carefully designed to meet the queries of intending investors, financial institutes, creditors, traders and growers engaged in dealing with horticultural crops. Further it will serve the purpose reference books to students, academicians and related people.

R.C. Upadhyaya

Plant Biotechnology and Horticultural Crops

All artificial manipulation made on a plant, or only on a part of a plant, could refer to "plant biotechnology". In our review, we shall separate the numerous plant biotechnology applications into two groups. The first group is related to tissue culture and go from a single cell, or a specific tissue, to the whole plant. These technologies are founded on a very old concept "the cellular totipotency", stated by Haberlandt (1902), but only demonstrated by Steward in 1958. His research team was the first to be able to transform a carrot cell line in some "artificial embryos, later called "somatic embryos"

The applications of the second group are more recent and derive from on excellent knowledge of the double helicoidal DNA structure proposed in 1953, by James Watson and Francis Crick. They base their strategy on the molecular biology science.

Horticultural Biotechnologies Related to Tissue Culture

The development of plant biotechnology has been related to the discovery of the growth regulators and their progressive introduction in tissue culture. In 1934, in the States, white was producing continuous growth of in vitro tomato roots. After that, went and Thiman's work on auxin, Gautheret in France (1935) and white in US (1938) were auxin, Gautheret in France (1935) and white in US (1938) were able to obtain division of isolated cells from Salix and tobacco, and finally to maintain undefined cell lines.

In 1957, Skoog and Miller demonstrated that 6-furfurylaminopurine, a cytokinin isolated from a DNA hydrolysate, could induce some caulogenesis, but also that the cytokinin/auxin balance could start different morphological programmes. A ratio higher than one unit induced caulogenesis, a ratio inferior at this unit induced roots. On this basis, a multibillion dollar industry was created. Hundreds of small and large nurseries and biotech labs throughout the world are propagating in vitro more than 1000 different plant species (Bajaj, 1991).

This micropropagation technique offers not only means for mass propagation, but also plays an important role to conserve elite or rare plants that are threatened with extinction. Moreover, meristem tip culture offers the possibility of virus eradication. These technologies started about 20 to 25 years ago, especially for ornamental plants, and for a few other crops: potato, strawberry, bananas ... Propagated clonally, it was very important to have a true to type propagation technique.

On the other hand, it's also possible to use tissue or cell culture to increase genetic variability. Undifferentiated cells obtained from callus, cells or protoplasts culture are produced and submitted to an selective pressure to improve a and submitted to an selective pressure to improve and fix the somaclonal variants.

Production of True to Type Plant

Meristem Tip Culture

Production of Virus-free Plants: In 1952 Morel and Martin were successful in regenerating a virus-free dahlia plant by the excision of some meristematic domes from virus infected shoots. Three years later, the same authors were able to eliminate virus A and Y from virus infected potato. Semal and Lepoivre (1992) reported that a virus-free sweet potato was producing 40T/ha in China by comparison of the 20 T/ha produced before meristem culture.

Virus eradication is dependent on several parameters. But to take advantage of the non uniform and imperfect virus distribution in the host plant body, the size of the excised meristem should be as small as possible. For Stone (1963), only tips between 0.2 and 0.5 mm most frequently produce virus free carnation plants. The explants smaller than 0.2 mm can't survive, and those larger than 0.7 produce plants that still contain mottle virus.

Today, no definitive explanation can be given to understand this virus eradication (Wang & Charles, 1991). various explanations have been given: absence of plasmodesm in the meristematic domes, competition between synthesis of nucleoproteins for cellular division and viral replication, inhibitor substances, absence of enzymes preiral replication, inhibitor substances, absence of enzymes present only in the cells of the meristematic zones, and suppression by excision of small meristematic domes. This last proposal could explain why some potato plant showing virus particles in the meristematic domes, could regenerate a virus free plant (Mellor & Stace Smith, 1977).

Shoot Tip Micrografting

When meristem tip culture fails, it is possible to graft small meristematic domes on young seedlings growing in vitro. In this way, Navarro et al. (1975) eradicated all the virus diseases from Spanish Citrus orchards. This technique was also very successful in eliminating virus diseases from peach trees (Mosella et al 1980).

Factors Influencing Virus Elimination

The success in virus elimination depends on the choice of the explants, the virus, temperature,....In general a high temperature (37C) is effective for virus diseases, but is depressive for viroids. The introduction of virus inhibitors (ribavirin) into the culture medium can't inactivate the viruses, but it prevents their replication. A combination of cytokinin and ribavirin, or a heat treatment in vitro, are sometimes used with success to inactivate some viruses difficult to eliminate by meristem culture (NRSV, PDV,..).

Micropropagation Techniques

During the Micropropagation Process, the Genetic Stabilit

During the micropropagation process, the genetic stability of the new shoots is dependent upon their origin. Axillary shoots issue from pre-existing buds and are normally true to type, indeed the meristematic cells are genetically very stable.

Adventitious shoots, such as somatic embryos, are neo-formed buds developed directly on some organs, or indirectly through a callus phase formed on this organ. So, if the mother plant presents a cell mosaic or chimaeric tissues, risks of genetic variation exists. It is similar in the case of an indirect regeneration, when the callus phase is too long.

Propagation by Axillary Shooting

This technique has proved to be the most applicable and reliable method of in vitro propagation. Axillary shoot growth is stimulated by overcoming apical meristem dominance.

Commercial tissue culture laboratories are now able to propagate a large number of herbaceous ornamental species and several woody plants in this way. However, the propagation of Pelargonium, Howea, and a few other horticultural plants are always difficult to propagate by axillary branching.

Propagation by Direct or Indirect Organogenesis

Adventitious shoots could arise directly from the tissue of explants without callus formation. Several plants of the family gesneriaceae (Saintpaulia, Streptocarpus,...) regenerate directly buds on leaf explants, likewise Lilium rnerate directly buds on leaf explants, likewise Lilium regenerates on scales.

However more often, like for Ficus lyrata, adventitious buds appear on callus. While coffee, cocoa trees, and many conifers are produced by somatic embryogenesis developed on callus or cell suspensions.

Constraints of in Vitro Micropropagation

The establishment of axenic culture could be difficult when the explants are coming from hot and humid countries. Moreover, in the course of micropropagation several authors observed a sudden appearance of endogeneous bacteria (Cassels, 1997). Nevertheless only Holland and Pollaco (1994) have demonstrated the presence of this kind of bacteria in more than 70 different species. For Leifert and Woodward (1997), the major source of contamination is the initial explant. They add that microbial contamination in commercial plant tissue culture laboratories is the most important cause of losses. For this reason, they support the introduction of a microbiological production control strategy.

A second important constraint during the initial phase mainly concerns woody plants. A mature tree must be rejuvenated to recover its morphogenetic competence. Franclet (1981) described various treatments, micrografts, cascade grafts, cytokinin treatments to reverse the adult phase to juvenile. The micropropagation of Acacia Senegal was only possible after rejuvenation by micrografting (Palmnly possible after rejuvenation by micrografting (Palma et al. 1997).

A distinction between chronological, ontogenic and physiological age (Monteuuis 1989) is needed to understand the rejuvenation concept. This was definitively demonstrated after a complete reversion of Sequoia sempervivens (Franclet & Franclet Mirvaux, 1992).

Improvement of Axillary Branching

The cost of micropropagated plantlets is also an important limitation of the techniques. In New Zealand, where they produce 2-3 million micropropagated radiata pine per annum, the relative cost of micropropagated planting stock had dropped from 13,8 times the cost of seedlings in 1988 to 6,9 by 1993 (Smith, 1997).

To reduce manpower costs, several improvements have been proposed. The more simple method was in vitro layering developed by Wang (1977) to clone PVX-free potato plants. The first plantlets placed on the medium in a horizontal position developed axillary shoots. They are harvested by cutting one centimetre above the medium surface, at 3 weeks intervals. A similar technique called 'hedging system ' by Aitken Christie and Jones (1987) was later used to produce Pinus radiata.

Ziv (1990) proposed for corm plants, Gladiolus and Nerine, a very rapid propagation system. She reduces the internodes and leaves by introduction of an anti-gibberellin agent in the medium. Finally, only aggregates of buds are formed, then they finally, only aggregates of buds are formed, then they are divided and introduced in bioreactors for mass production. Similar systems were developed in Gembloux, to propagate carub trees and asparagus.

Since 1988, Duhem was producing very large quantities of Eucalyptus plantlets in Petri dishes without anti-gibberellin but in complete darkness. Transfers from one Petri to another is made by a simple squashing (Boxus et al, 1991).

Somatic Embryogenesis Propagation

For genetically stable species, somatic embryogenesis offers a very fast scaling-up system, especially when it's possible to produce embryos in bioreactors. Unfortunately this production via fermentors was not as simple as first envisaged. And today, only a few model plants are successfully produced by such technology: carrot, celery. Other applications remain at the experimental stage: coffee, oil, palms, conifers, Euphorbia pulcherrima,... and several other horticultural species.

Several bottlenecks limit the use of this interesting technology. One of the main problems is genetic stability. So, despite the clonal nature of nucellar embryos, different morphological anomalies can occur among mango somatic embryos, as it was also observed in Citrus plants derived from nucellar cultures (Litz et al., 1993).

Another difficulty is the loss of embryogenic capacity over time, a phenomenon observed with different species. It is also a phenomenon observed with different species. It is also important to note that somatic embryogenic lines of conifers are always originated from immature embryos.

Artificial Seeds

Another very interesting possibility of the somatic embryogenesis technology has been developed during the past fifteen years by Redenbaugh and his team (1991, 1993). They were able to encapsulate somatic embryos by hydrogel coatings (sodium alginate), producing single embryo artificial seeds. To date, some improvements offer the possibility to directly plant the artificial seeds in the greenhouse on special substrates (vermiculite, sans,...). This methodology will provide in the future a good technique to reduce the cost of transplants.

Gene Bank

Tissue culture methods offer the opportunity for in vitro collecting, rapid multiplication and distribution of important, elite, or rare plants that are threatened with extinction. The two major in vitro storage strategies are slow growth and cryopreservation. Since the first results of Seibert (1976), who was able to initiate shoots from carnation shoot apices frozen to -196C, this technique is now successful for many of horticultural species. Dereuddre et al. (1991) have provided a very simple technology to freeze encapsulated meristems in dried alginate beads. It works for pear, strawberry, eucalyptus, potato,....

Some Ior Pear, Strawberry, Eucalyptus, Potato

Some international institutions have very large collections of old and current varieties, available for exchange and introduction in crop improvement programmes. The International Potato Centre (CIP) in Lima, Peru, has a large world potato collection. Germplasm of sweet potato and cassava is at the International Institute of Tropical Agriculture (IITA), Ibadan, Nigeria.

The movement of germplasm involves the risks of accidentally introducing plant quarantine pests along with the host plant material. To limit these risks, the plant material should be transferred from one country to another as in vitro cultures through a transit Centre, where it should be indexed. For bananas, in the framework of INIBAP, the transit Centre is the Catholic University of Leuven in Belgium where a very large in vitro germplasm exists.

Tissue Culture Techniques

Somaclonal Variations

The production of plantlets by callus regeneration, cell suspensions, protoplast cultures, could present some deviations with regard to the mother plant. This is a way to increase the genetic variability. Associated with a selective pressure (stress to toxins, pH, salinity, cold,...) it's used to obtain resistant lines. Indeed, after regeneration, plants can express new potentialities rarely obtained another way. Stable and profitable variants are sale another way. Stable and profitable variants are selected and introduced in breeding programmes. In 1976, a Pelargonium cv Velvet Rose was created by this technique (Reisch, 1983) Later, other applications were described, as in sugarcane, tomato, red pepper,... Following Sibi (1994), extrachromosomic systems must be involved to explain the genetic behaviour of some new expressed tissue culture characteristics. She uses the term "epigenic" to evoke a whole of hereditary elements which belongs to the cytoplasmic or nuclear compartments, but are not transmitted by mendelian rules.

Haplodiploidisation

Many species are able to produce haploids through different in vitro techniques. The oldest was anther culture of androgenesis. To date, the results vary considerably from one species to another. Solanae (datura, tobacco, red pepper, eggplant, petunia), cereals (wheat, barley, rice, triticale, maize) or crucifers (soybean, cabbage) are species easy to regenerate by anther culture. To the contrary, tomatoes, leguminous, or Compositae are recalcitrant.

Ovule culture, or gynogenesis, was a successful technique for Gerbera, beet, courgette,... However, the most common technique is pollination with irradiated pollen, in order to induce in vivo parthenogenesis. The oospheres developed in embryos without fertilisation are saved by embryo rescue.

The haploid plantlets (n chromosomes) are tree. The haploid plantlets (n chromosomes) are treated with colchicine to produce fertile homozygous lines, called "double haploid lines". This haplodiploidisation technique could give immediately new elite genotypes, hybrid parents after in vitro propagation, as asparagus supermales (MM), or useful genetic material to establish gene mapping. A good synthesis book concerning all these possibilities was published by Bajaj (1990).

Protoplast Culture

Protoplasts are the smallest units able to regenerate a whole plant. Therefore protoplasts cultures can serve to enlarge genetic variability by introducing somaclonal variation.

However, the main interest of protoplasts, is their capacity to fuse and to produce hybrids or cybrids, new organisms most often unknown in nature. Indeed, protoplasts allow to transgress the barrier of the botanical species or genus. Naked protoplasts can accept without rejection external elements: nuclei, cytoplasmic corganelles, liposomes,.... containing genetic information. The first protoplast fusion application was a cytoplasmic transfer from one genotype to another to induce male sterility (CMS) from mitochondrial origin. These male sterile hybrids are interesting to produce F1 hybrids (Brassica, Cichorium,...).

Another cytoplasmic transfer concerns the introduction of chloroplastic DNA to induce herbicide (atrazine) resistance, from Solanum nigrun duce herbicide (atrazine) resistance, from Solanum nigrun to tomato for instance (Jain et al. 1988).

Many intergeneric and interspecific hybridations are possible but not always profitable because these hybrids are too often infertile. This problem does not exist if we can vegetatively propagate the hybrids, as in the case of Citrus rootstocks. In Florida, Grosser (1990) was able to analyse several thousand Citrus hybrids.

Contributions of Molecular Biology in Horticulture

More and more pathologists are using molecular probes for early detection of diseases. However, the main contributions of modern biotechnology remain probably in the hands of the plant breeders. Indeed, molecular markers are of great use in detecting desirable new genes, or to identify important QTL. This marker-assisted selection is very useful for many vegetable plants. Other applications are DNA - fingerprinting and genetic engineering. This recent technology aims to insert and to express new specific genes into a selected plants.

Marker-assisted Selection

Molecular markers and genome mapping will reach a large extension in the next breeding programmes of tomatoes, peas, cabbages, melons, potatoes,...

It will be possible to improve the speed and the incorporation efficiency of desirable new genes by utilizing of closely linked selectable molecular markers genes by utilizing of closely linked selectable molecular markers. The number of backcross will be reduced. Tanksley et al. (1981) show that only with 12 markers, one on each tomato chromosome, the composition of the recurrent parent genome is similar to that observed in the absence of selection after the third backcross two years later. Therefore, it's easy to understand the high improvement that we can get with the 700 tomato markers already known in 1990 for this crop (De Verna and Paterson, 1990).

The utilization of molecular markers could also identify important quantitative trait loci (QTL). Recently Foolad and Chen (1998) have identified 13 RAPD markers at eight genomic regions, that were associated with QTLs affecting salt tolerance during germination in tomato.

DNA Fingerprinting

The DNA polymorphism observed by RFLP, AFLP, or RAPD, allows cultivar identification of fruit trees, apple, citrus,... or other vegetatively propagated plants. Although some mutants, as coloured fruits, escape these identification techniques, some people would evaluate the genetic uniformity of regenerated plants in this way.

Genetic Engineering

Genetic engineering consists of introducing a foreign gene into a plant genome to create a new function. Or at the contrary, it could be to reduce or suppress an existing gene function, as in the antisense strategy.

1. *Historic:* In this field, progress were very rapid. In 1974 the pathogenicity of Agrobacterium tumefaciens, due to a large plasmid called Ti, was demonstrated. In 1977, Chilton et al. demonstrate a stable incorporation of plasmid Ti-DNA into higher plant cells. In 1983, the team of Van Montagu and Schell in Gent succeeded in producing a transgenic plant, and eleven years later the first transgenic cultivars were available in the market.
2. *Transgenic Plants with New Agronomic Traits:* Two different techniques are commonly used to introduce genetic information (DNA) inside cells, protected by

pectocellulosic walls. A particle bombardment process, called biolistic, is used with success for monocots gynmosperm, squash, peas,...Today, this technique allows the bombardment of meristems and avoids the difficulties of regeneration. But the mediation of Agrobacterium remains till now the most routinely system to transform dicotyledonous plants.

The main agronomic traits introduced in horticultural plants and already commercialised are Bt toxin and herbicide resistance. Other studies concern virus resistance, male sterility...

3. *Antisense Strategy:* Calgene created in 1994 the first commercial transgenic plant, a long shelf life tomato, by the suppression of polygalacturonase activity due to an antisense gene (Smith, 1988). However, t activity due to an antisense gene (Smith, 1988). However, this Flavr Savr tomato variety was removed from the trade 3 years later, because of its disease susceptibility and its lack of productivity !

   Later, other tomato varieties with long storage qualities were obtained by the utilization of an antisense RNA inhibition of ACC synthase or ACC oxydase, two ethylene precursors.

   The antisense technique was also used to reduce the lignification of woody plants, by blocking the enzymes involved in the precursor of lignin biosynthesis. Another interesting application was the induction of white flowers in petunia and in different other ornamental plants by the suppression of chalcone synthase activity.

4. *Transmission of the New Traits:* These traits induced by the transferred DNA are transmitted to the progeny as a dominant Mendelian character. Nevertheless in some cases, transgenic plants with a strong gene expression can generate progenies with only a faint no gene expression. This problem is not clearly elucidate.

5. *Marker, Reporter, Promotor and Expression Genes:* To select the transformed cells, some marker genes, very often resistant to antibiotics or to herbicides, are attached

to the coding sequence, as a promoter or an expression gene. Those allow the new gene to express in the whole plant or in a specific plant tissue.

Some reporter genes are used to follow the evolution one.

Some reporter genes are used to follow the evolution of the transformed cells. The most frequent reporter genes are the gene "gus" or the gene lux (luciferase).

6. *Horticultural Transgenic Crops already released in USA:* The Flavr Savr tomato was the first genetically engineered whole food approved for commercial sale in 1994. Four other transgenic tomatoes with a delayed ripening were approved later (1995 and 1996).

In 1995, potatoes with Bt genes and a squash cultivar resistant to two viruses were released. Two years later, another squash cultivar resistant to 3 viruses and a papaya line also resistant to viruses were approved. Two other horticultural crops are waiting the authorities approval: red hearted cichory (Radicchio) with male sterility and resistance to herbicide, and a tomato with Bt toxin.

It there are so few commercial transgenic plants in horticulture, it's probably due to the relatively low acreage of these horticultural crops in comparison with other agricultural crops. In 1997, in the world 5 million ha were planted with glyphosate resistant soybeans released by Monsanto alone.

Nowadays, it's true also that many transgenic plants exist in research laboratories, awaiting authorization for field testing, as they are very interesting model plants for learning physiology: lettuce with less nitrate by increasing nitrate reductase gene active with less nitrate by increasing nitrate reductase gene activity, a yeast ribonuclease gene to reduce viroid infection on potato, modification of lignin synthesis in plant to produce trees adapted to paper industry, or timber, or biomases production, regulation genes to control tree architecture,....

Situation of Biotechnologies in the World: Development and Risks

Situation in the Young Countries or in Developing Countries: A general use of in vitro micropropagation in the developing countries is now a reality, as it was predicted by Albert Sasson (1993). African countries for instance are producing potatoe, banana, cassava...

Even the production of transgenic plants is no longer a prerogative of the Northern countries. Different experiments are now starting in the South, specially in the international laboratories from CGIAR. Although CGIAR's expenditures on biotechnology for 1997 raise only 24.2 million US$ on an annual budget of 345 million US$ (Biotechnology and Development Monitor, December 1997, 33, pp12-17) this Consultative Group on International Agricultural Research plays a very important role as contributor to agricultural research for developing countries.

The large international research centres IITA (Ibadan, Nigeria), CIP (Lima, Peru), CIAT (Cali, Columbia),... belong to the CGIAR. Their roles and NGO's are primordial. The genetic transformation their roles and NGO's are primordial. The genetic transformation of cassava is an excellent example. Its tuberous roots provide food for over 500 million people, mostly small-scale farmers. Unfortunately, till recently this integral plant for food security in developing countries has been recalcitrant to transformation approaches. However, thanks to the participation of international institutes located in cassava growing countries, CIAT and IITA with the help of four development associations from Swiss, UK, The Netherlands and USA, were able to initiate procotols for cassava transformation and regeneration, (Biotechnology and Development Monitor, March 1997, 30, pp. 16-18).

Other networks, like the Cassava Biotechnology Network (CBN) previously described, exist. REDBIO, a technical cooperation network on plant biotechnology, is supported by FAO to promote a best use of scarce manpower, equipment and other resources in Latin America.

On the other hand, more and more private laboratories exist in the developing countries which have a high scientific and technical level (China, Singapore, Taiwan, India, Brasil, Mexico, Chile,...). In these countries, European, American or Japanese agrochemical companies are investing. Also these countries will have their chance to develop their own technologies, and to export to the Northern markets.

Meanwhile, genetic engineering techniques are being applied mainly to crops which are important for g applied mainly to crops which are important for the industrialized world, not crops on which the world's hungry depend. Therefore, it is unrealistic when Monsanto writes "the experts said that biotechnological innovation will increase the crop productivity without occupation of new lands, saving tropical forest of best quality and animal habitat". However, the World Bank was promising in a 1997 publication Bioengineering of crops that" transgenic crops could improve food yield by up to 25 per cent in the developing countries and could help to feed an estimated additional three billion people over the next 30 years".

But, not all people are confident in this prospective. To the contrary, some are afraid that the poorest countries, where more than 700 million people are chronically undernourished will miss the biotechnological revolution.

Nevertheless, there is some hope to transfer to the developing countries high technology, which don't undermine the environmental and social network. So, the Agricultural Biotechnology for Sustainable Productivity Project (ABSP project) is managed by Michigan University. Four American universities, two research centres from France and USA, and two American private companies are also involved in this project. The ABSP's objectives are the reduction of losses due to pathogens and the reduction of losses due to pathogens and pests by using transgenic plants (sweet potatoes, potatoes, cucurbits, tomatoes) and by cloning commercial value-plants (bananas, pineapple, coffee).

Situation in the Northern Countries

1. *Mass Propagation:* In 1988, Pierik (1991) predicted a European in vitro propagation of 212 million plantlets/ year. He was also predicting an important development of this activity for the future. Following Pierik this quantity accounted for only 5 % of the plants able to be propagated by vegetative means. However a few years

later Pierik's prediction was not confirmed. The world production was estimated at 600 million. In USA, Zimmerman (1997) was recording a production of 121 million plantlets for a total of 110 laboratories, half produced in 11 laboratories, and 6 labs were propagating each more than 6 million plants per year.

In Europe, only the Dutch production of micropropagated plants is precisely known. In 1990 the total production was 95.5 million. In 1995, the Dutch in country production reached only 53,7 million, whereas importations increased 77,3 million. (37 million from Poland and 17,1 million from India) (Pierik, data unpublished). These laboratory relocations to low-cost manpower countries are commonly observed, and put the question of the quality management.

2. *Transgenics:* The GMOs invade progressively all the US lands: 1.6 % in 1996, 15 % in 1997, 50 % in 2000..... It will represent an estimated market of 3,9 million US$ in 2003 ! The largestt agrochemical companies are merging to become giants of agrobusiness such as Novartis, Monsanto, Zeneca,...

   Following Monsanto in 1996, about 950 T insecticides were saved by introduction of resistant cotton, id est a net savings of 81 US$/ha for the grower. Glyphosate-resistant soybeans (4 million ha in 1997) allow a savomgs of 44 to 49 US$/ha. (Information from cultivar, suppl. n 436, Feb. 16, pp. 24-37, 1998).

3. *Perspectives, Limitations and Environmental Risks*: Ecological impact related with the introduction of GMOs is always an open debate.

   The application to agriculture of these new technologies certainly opens interesting perspectives, but also raises potential problems. The risk of crop transgene spreading has been demonstrated. A researcher of Clemson University in South Carolina reported "that in a population of wild strawberries growing within 50 meters of a strawberry field, more than 50 % of the wild plants contained marker genes from the cultivated strawberries" (Kling J. 1996).

A Danish team (Mikkelsen et al, 1996) have shown a possible rapid spread of genes from oilseed rape to the weedy relative *Brassica campestris*.

There are other risks. *Brassica campestris*.

There are other risks. The introduction of Bt gene allows a drastic reduction in the use of toxic chemicals for crop protection. But a poorly controlled use of Bt-technology can destroy more effectively the predators than the pests. Or when" many crop plants are transformed with similar effective traits, in such situation, many polyphagous pest species, which by nature are more flexible evolutionarily than those that have a narrower diet, are likely to overcome Bt resistance very quickly". (Hokkanen, 1998).

Therefore, before releasing a transgenic plant, any risk has to be weighed against the benefit of the transgenic crops. We must not forget that annually in the world 500.000 acute pesticide poisonings, with 5000 deaths, are observed.

Estimated Global Economic, Environmental, and Human Health Benefits of Bt Transgenic Plants. (following Hokkanen, 1998)

Global Market Penetration of Transgenic Plants (% use)	Reduced Use of Conventional Insecticides (US$ millions)	Cost of Transgenic Plant (US$ millioTH=112> Cost of Transgenic Plant (US$ millions)	Economic Gain (US$ millions)	Estimated Environmental Gain (US$ millions)	Estimated Human Health Gain (US$ millions)
1	90	45	45	91	16
10	900	47	853	917	157
25	2250	49	2201	2293	394
50	4500	51	4449	4585	787

Horticulture Research in India? Infrastructure, Achievements, Impact, Needs and Expectations

India has a wide variety of climate and soil on which a large range of horticultural crops such as, fruits; vegetables, potato and other tropical tuber crops; ornamental, medicinal and aromatic plants; plantation crops; spices, cashew and cocoa are grown. After attaining independence in 1947, major emphasis was laid on achieving self sufficiency in food production. Development of high yielding wheat varieties and high production technologies and their adoption in areas of assured irrigation paved the way towards food security ushering in green revolution in the sixties. It, however, gradually became clear that horticultural crops for which the Indian topography and agro climates are well suited is an ideal method of achieving sustainability of small holdings, increasing employment, improving environment, providing an enormous export potential and above all achieving nutritional security. As a result, due emphasis on diversification to horticultural crops was given only during the last one decade.

Research Infrastructure

The Indian Council of Agricultural Research is the premier agency which pioneered systematic research on agricultural crops in the country. Horticulture research in India received very little attention till the 3rd Five Year Plan. The establishment of the Indian Institute of Horticultural Research at Bangalore and starting of eight All India Coordinated Crop Improvement Projects to cover different horticultural crops was a landmark in the history of horticulture in 4th Five Year Plan (1969-74). Rapid expansion of infrastructure took place in 7th and 8th Plans. Today, the horticultural research in the country is being carried out at eight ICAR institutes (with 26 regional stations), 10 National Research Centres (on major crops) and a Project Directorate on Vegetable crops. Area specific, multi-disciplinary research is also being conducted under 14 All India Coordinated Research Projects each on Tropical, Subtropical and Arid Fruits; Vegetables, Potato, Tuber Crops and Mushrooms; Ornamental Crops, Medicinal and Aromatic crops; Palms, Cashew, Spices and Betel vine; and Post Harvest Technology at 215 centres located at various research Institutes, and State Agricultural Universities. In addition, four net work projects each on hybrid research in vegetable crops, drip, irrigation in perennial horticultural crops, protected cultivation of ornamental crops and *Phytophthora* diseases of horticulture crops are now in operation. Research on horticulture is also being undertaken at several multi-crop, multi-disciplinary Institutes. Departments of Horticulture in 24 Agricultural Universities, one deemed to be University and one full fledged University of Horticulture and Forestry are also engaged in horticultural research. Besides 280 adhoc schemes supported from Agriculture Produce Cess Fund and a number of foreign-aided projects have also been in operation on specific problems of different horticulture crops. As a result, the country now has a sound research infrastructure in horticulture to meet the growing needs and expectations of the fast developing horticulture industry.

Budgetary Support

The investment in horticulture research by the ICAR in the Central sector has increased significantly in the last two Plans. The Plan allocation for horticultural crops started in 4th Plan (1969-74) with a modest allocation of Rs. 34.78 million and was enhanced to Rs. 319.56 million in the 7th Plan (1985-90) and to Rs. 1047 million in the 8th Plan (1992-97). Non-Plan expenditure also increased from Rs. 73.55 million in the 5th Plan to Rs. 768 million in 8th Plan. Overall increase in Plan investment in 25 years has been of the order of 2775.21 per cent. The per cent budget allocation for horticulture research out of the total budget for agriculture research rose from 6.1 in 5th to 6.5, 6.67 and 7.7 in 6th, 7th and 8th five year plans, respectively. Similarly, expenditure for Central Sector Schemes of the Department of Agriculture & Cooperation for horticulture crop development also rose tremendously from Rs. 20.5 million (4th Plan) to Rs. 76.18 million (5th Plan), Rs. 146.37 million (6th Plan), Rs. 250 million (7th Plan) and Rs 10,000 million (8th Five Years Plan).

Budgetary Support (Million Rupees)

Five year plans	*Total for Agriculture Research*	*Share of Horticulture (%)*	*Total Development Support*
IV 1969-74	6105	N.A	20.50
V 1974-78	7292	6.10	76.18
VI 1980-85	10684	6.50	146.37
VII 1985-90	8445	6.67	250.00
VIII 1992-97	15165	7.70	10,000.00

Manpower

Nearly one sixth of the total strength of 5906 scientists working in ICAR is allocated for horticulture research in ICAR Institutes. Besides, approximately, 560 scientists are working in State Agricultural Universities in ICAR funded All India Coordinated Projects. In addition, a large number of scientists are working on horticultural crops in State Agricultural Universities.

Total Scientists in ICAR	5906
Scientists working in the Division of Horticulture	832
Scientists working in other ICAR Institutes	150
% of Total	15
Scientists working in SAUs in ICAR funded projects	560

Research Achievements

Introduction and Cultivation of New Crops: Several new crops have been introduced for commercial cultivation, *e.g.*:

- Kiwi fruit in sub-mountain areas of North India
- Olive in mid hills of North Western Himalayas
- Low chilling stone fruits in the North Western plains
- Oilpalm in coastal states of Karnataka, Andhra Pradesh, *etc.*
- Gherkin in south and west India
- Baby corn and sweet corn in certain specific pockets, and
- Broccoli, Brussels' sprouts, asparagus, celery, parsley near the cities.

Crop Improvement

- A large number of high yielding varieties developed in several horticultural crops *e.g.* fruits (76), vegetables (160), potato (29), other tuber crops (24), ornamental crops (300), palms (20), spices (51), cashew (33) and betel vine (1).
- First seedless variety of mango developed.
- 40 F_1 hybrids developed in brinjal, tomato, chillies, cauliflower, carrot, capsicum and muskmelon.
- Self incompatible lines in cauliflower, gynodioecious lines in cucumber and muskmelon, genetic male sterile lines in tomato and temperature tolerant strains of button mushroom developed.

Propagation of Quality Planting Material

- Standardized propagation technique for many fruits hitherto propagated by seed. *e.g., aonla, bael, ber,* black pepper, cardamom, cashew, cassia, cinnamon, clove, custard apple, jack fruit, *jamun,* nutmeg, sapota and walnut.

- Standardized Seed Plot Technique resulting in successful disease free potato seed production in the tropics and sub tropics of the country. Standardized method of micro-propagation and *in vitro* micro-tuber production in potato.
- Identification of suitable parental lines for production of True Potato Seed (TPS) and Standardized technology for raising commercial crops.
- Micropropagation protocols developed in banana, black pepper, betel vine, cardamom, ginger and turmeric.
- Production of coconut hybrids through establishment of Seed gardens of Tall (T) x Dwarf (D) and D x T hybrids.
- Standardized rootstocks in citrus, grape and apple.

Agrotechniques

- Standardized high density plantations in banana, citrus, mango and pineapple and high production technology in several crops *e.g.,* pineapple, black pepper and cardamom.
- Year round production technology in tomato and "off season" cultivation of onion and cauliflower developed.
- Arecanut, coconut and potato based cropping systems developed to maximize productivity under high management conditions.
- Standardized use of several plant growth regulators and chemicals now commercially employed in production and quality improvement of horticultural crops *e.g.,* paclobutrazol for induction of flowering in mango; gibberellic acid for improving berry size and quality in grape; Maleic hydrazide for preventing sprouting in onion and potato; Dormex for hastening bud burst in grapes; and Boron and Calcium for changing flower cycle in some cucurbits.

Crop Protection

- Developed improved disease detection techniques such as ELISA and ISEM for improving seed quality and tissue culture technique for rapid multiplication of potato.

- Developed IPM for fruit borer in brinjal, diamond back moth in cabbage, thrips in chillies, phytophthora foot rot in black pepper, "Katte", rhizome rot in cardamom, rhizome rot in ginger, late blight and bacterial wilt in potato.
- Developed biological control measures for mealy bug in grape, fruit borer in tomato and okra.

Post Harvest Management

- Preharvest treatments to control post harvest losses in citrus, mango and grape standardized.
- Maturity standards for mango, guava, grape, litchi and ber standardized.
- Chemical treatment for regulation of ripening in mango, sapota and banana standardized.
- Optimum storage temperatures worked out for several fruits, vegetables and tuber crops.
- A mango harvester, fruit peeler, hand and pedal operator cassava chipping machine, harvesting tools (5-14 times efficient); Implements for mechanization of potato cultivation *e.g.*, oscillating tray type potato grader, fertilizer application cum line marker, potato culti-ridger, soil crust breakers, potato digger and automatic potato planter/diggers developed.
- Low cost environment friendly storage system for fruits, vegetables, potato and onion developed.

Impact of Research

Fruits: Area under fruits increased from 1.22 million hectares to 3.35 million hectares in 1995-96. India with a production of 41.50 million tonnes (1996-97) is the second largest fruit producer next to China (45.46 million tonnes) with a share of 8% in world fruit production.

India produces 65% and 11% of worlds mango and banana, respectively, ranking first in the production of both the crops. It has the highest productivity in grape in the world. Significant expansion has taken up in apple, *aonla, ber,* pomegranate and sapota cultivation.

Vegetables: India ranks second in the world vegetable production after China (71.59 million tonnes). Vegetable production has increased three times during the last 50 years. A large area is now covered with F_1 hybrids in vegetable crops resulting in increased yield and better socio-economic status of farmers. Vegetables like tomato, cabbage, cauliflower, radish and onion are now produced almost round the year. India has attained self sufficiency in seed production of temperate vegetables.

Mushroom: Mushroom cultivation has spread to almost all parts of the country. Its production has increased from 100 tonnes in 1970 to 30,000 tonnes in 1996-97. The productivity has increased from 10-12 kg m^{-2} in 1985 to 18-22 kg m^{-2} in 1995.

Potato: In potato, area, production and productivity has increased from 0.234 million hectares, 154 million tonnes and 6.59 t/ha in 1949-50 to 1.14 million hectares, 1924 million tonnes and 16.9 t/ha respectively. This increase is 12, 4.9 and 2.6 times, respectively. The annual compound growth rate for potato during this period was 6.07 compared to 5.6% for wheat, 2.7% for rice and 2.74 % for total food grains. India is the only country in South East Asia having a national disease free seed production programme producing 2600 tonnes of breeder's seed annually.

Production, Demand and Projections of Horticultural Crops

	1996-97		Demand		Target 2002	
	Area*	Production**	Area	Production	Area	Production
Fruits	4.54	46.97	5.24	59.47	4.93	56.00
Vegetables	5.12	80.80	6.96	131.2	5.73	108.00
Spices	2.54	2.78	2.94	4.43	2.60	3.90
Coconut	2.00	9.75	2.67	15.60	2.57	15.00
Cashew	0.63	6.45	0.69	0.80	0.60	0.70
Arecanut	0.24	0.31	0.24	0.39	0.25	0.39
Total	15.07	141.06	18.74	211.89	16.68	183.99

**Area in million hectares;*

***Production in million tonnes*

Cassava: In cassava, productivity has increased from 7t/ha in 1960-61 to 22t/ha during 1992 which is more than double the world average (9.81t/ha). Sago and starch industry based on cassava has been developed.

Floriculture: Protected cultivation of cut flowers started a decade back and India has already entered the world cut flower market. Micro propagated ornamental foliage plants are being exported in millions internationally. Export of dried flowers from India is increasing.

Coconut: In coconut, area has increased from 1 million hectares in 1980 to 1.793 million hectares in 1996-97. India has become one of the largest coconut producing country of the world. Production of coconut has gone up from 5677 million nuts to 13968 million nuts. The productivity in coconut has increased from 5249 nuts/ha to 7808 nuts/ha. Coconut contributes 700 billion rupees to the GDP of the country. The contribution of the crop to the total edible oil pool in India is around 6 per cent. India also exports coir and coir products derived from coconut husk to the tune of 2260 million Rupees.

Export of Horticultural Produce

Arecanut: India continues to dominate the world in area, production and productivity of arecanut and has achieved self sufficiency in arecanut production (0.27 million tonnes). Most of the production is domestically consumed.

Oilpalm: Area under oil palm has gone up from 200 ha in 1965 to 40,700 ha in 1996-97. Average productivity in oil palm plantations is now 4-5 tonnes/ha which compares favourably with other countries.

Spices: India is the largest producer (2.48 million tonnes), exporter (0.20 million tonnes) and consumer of spices. Indian spices flavour foods in over 134 countries. Spice exports touched Rs. 11800 million during the last year.

Cashewnut: In cashewnut, area has increased from 0.176 million hectares in 1961 to 0.659 million hectares in 1996-97. The production in cashew has gone up from 0.079 million tonnes to 0.430 million tonnes in 1996-97. India exported cashew kernels worth Rs. 13000 million (362 million $US during 1996-97). Export of cashew rising @ 27% per annum. These export earnings are exceeded only by coffee and rice among agri-exports.

Needs and Expectations

In spite of significant achievements in horticulture R&D, a number of challenges still need to be met. These are:

- Inadequate supply of quality planting material,
- Heavy losses caused by several biotic and abiotic stresses, and
- Several unresolved chronic disorders.

As a result, the productivity per unit area is low, resulting in high cost of production. Further, the quality of produce in many cases is far from satisfactory. The post harvest losses continue to be high. Full advantage has yet to be taken of several frontier areas *e.g.,* bioteohnology, protected cultivation, computer aided management of inputs, integrated nutrient management, leaf nutrient standards, biofertilizers, integrated pest management and mycorrhiza. There is also need for change both in the content and approach of research which can be taken up in partnership with private sector on aspects like production of hybrids, green house production of flowers, biotechnology, value addition and export. The future growth of horticulture industry will largely depend on new and globally competitive technologies. As such, ambitious research programme is called for in horticultural crops in the following thrust areas.

Genetic Resources

- Introduce fruits like mangosteen, durian, rambutan, longan, macadamia and berries not yet commercially exploited in India.
- Widen genetic base in mango (*Mangifera* species of South East Asia), Citrus (newly developed rootstocks), apple (scab resistant cultivars), guava (coloured varieties), papaya (species and varieties) and litchi (varieties from South East Asia and China). Also, widen genetic base for improvement of medicinal and aromatic plants required by pharmaceutical industries.
- Enrich germplasm collections with species/ cultivars which are high yielding and resistant/tolerant to different biotic and abiotic stresses.

- Exploit full potential for cultivation of avocado, kiwi and olive.
- Standardize long term techniques for cryopreservation of propagating materials and pollen grains.
- *In situ* conservation of endangered genetic wealth.

Crop Improvement

- Develop dwarf rootstocks/ scion varieties for high density planting and export in mango, litchi, sapota, citrus, *ber*, coconut, arecanut, oilpalm and cashewnut.
- Induce and exploit useful genetic changes through mutations in commercial cultivars.
- Develop rootstocks and scion varieties in fruits resistant/ tolerant to major biotic and abiotic stresses, *e.g.* malformation in mango, guava rootstocks for wilt, citrus rootstocks against phytophthora, root wilt resistance in coconut, frost and PRV resistance in papaya, weevil in sweet potato, cassava mosaic and stress tolerance in arecanut and spices.
- Develop F_1 hybrids in vegetable crops for yield, nutritional quality and export.
- Development of virus resistant potato varieties having high tuber dry matter and low accumulation of sugars during low temperature storage and varieties with high temperature tolerance.
- Evolving varieties with bold nuts and higher shelling percentage in cashew.
- Domestication of indigenous medicinal plants with bulk demand both in the modern and traditional systems of medicine.
- Studies on the quality control, seed health, packing and storage of seeds of annual crops.

Crop Production

- Standardise rootstocks for all important fruit crops *e.g.* mango, guava, litchi, sapota, *ber* and walnut.
- Develop horticultural crop based cropping systems for different agroecological regions.

- Develop integrated nutrient management system *i.e.* efficient utilization of chemical fertilizers, use of bio-fertilizers and organic material using leaf nutrient standards.
- Standardize water management practices in major crops including micro-irrigation and fertigation.
- Develop techniques of organic farming for export oriented horticultural crops.
- Standardize practices for production of horticulture crops particularly flowers and vegetables under protected cultivation involving computer aided supply of inputs.
- Standardize production technology for quality crops for export *e.g.* mango, grape, litchi, cashew, potato and spices.

Crop Protection

- Develop IPM strategies for important pests of commercial crops.
- Develop biological control for important diseases and insects affecting commercial crop production.

Post Harvest Management

- Reduce losses occurring during harvesting, storage and transport and improve shelf life of perishable horticultural commodities.
- Conduct basic and applied research in CA/ modified atmospheric storage of high value perishable crops.
- Improve indigenous low cost storage systems developed for onion, potato, fruits and vegetable crops.
- Develop techniques for bulk preservation of fruit pulps, improve drying systems for raisins, mushroom and different vegetables, and standardize technique for frozen vegetables.
- Initiate post harvest research on ornamentals with special emphasis on export market.
- Utilize wastes for development of economically viable products.
- Develop/fabricate low energy requiring machinery for horticultural crop production.

Biotechnology

- To standardise *in vitro* culture techniques for mass multiplication of rootstock/scion of difficult to micro propagate plants like mango, guava, sapota, litchi, walnut, date palm, coconut and oil palm.
- Development of protocols for rapid propagation of seedless water melon; vegetable crops with male sterility and self incompatibility barriers *e.g.,* tomato and capsicum and selected ornamental crops.
- Use of anther culture system for production of di-haploids for integration with breeding programmes.
- To identify molecular markers based on RAPD and RFLP in important varieties of horticultural crops.
- Develop transgenic plants with endogenous resistance to insect pests *e.g.* bacterial canker in kagzi lime, salt and drought tolerance in tomato and capsicum.

National Problems

Intensify interdisciplinary research to find solutions to nationally important problems *e.g.*, malformation and irregular bearing in mango, wilt in guava, die back in citrus and root wilt in coconut.

India has a good natural resource base, an adequate R&D infrastructure and excellence in several areas. As a result, the horticultural scenario of the country has been changing fast. Both production and productivity of several crops has increased manifold and India can boast itself as a leading horticultural country of the world. Many new crops have been introduced and many others have adapted to non conventional areas. Some other crops are under adaptive trials. A number of turn key projects in mushrooms and flower production have been established. Near self-sufficiency has been achieved in many crops. Export of fresh as well as processed fruits has been increasing. The demand of horticulture produce is on the rise due to increasing population, changing food habits, realisation of high nutritional value of horticultural crops and greater emphasis on value addition and export. However, several challenges are yet to be met. These are, fast eroding genepool, fast population build up, shrinking land and other natural resources, serious production constraints, both biotic and abiotic and huge post harvest losses. Further, in the era of globalisation, produce has to be of international quality and globally competitive. The future expansion of horticulture has to be in and semiarid areas and on under utilized crops.

While the impact of green revolution in India was felt mainly in assured irrigation areas, horticultural crop production has brought prosperity even in and semiarid areas. Horticulture is no longer a leisurely avocation and is fast assuming position of a vibrant commercial venture. Nature has placed India in a state of advantage and it is now on us horticulturists to work towards ushering in a GOLDEN REVOLUTION in years to come in India.

Research Preparedness for Accelerated Growth of Horticulture in India

The horticulture scenario of the country is rapidly changing. The production and productivity of horticultural crops have increased manifold. Production of fruits and vegetables has tripled in the last 50 years. The productivity has gone up by three times in banana and by 2.5 times in potato. Today horticultural crops cover about 25 per cent of total agricultural exports of the country. The corporate sector is also showing greater interest in horticulture. A major shift in consumption pattern of fresh and processed fruits and vegetables is expected in the coming century. There will be greater technology adoption both in traditional horticultural enterprise as well as in commercial horticulture sectors. Diversification and value addition will be the key words in the Indian horticulture in the 21st Century.

Horticulture research in India is about four decades old. Systematic research on fruit, vegetable and ornamental crops began in 1954 with the initiation of independent institutions and programmes. The research agenda is designed relevant to national plans and priorities for the horticulture development. Today, eight ICAR institutes with 27 regional stations, 1 project directorate, 10 national research centres, 16 all India coordinated research projects (AICRPS) with 223 research stations, 1 full-fledged university of horticulture, 25 state agricultural universities and 7 multi-disciplinary institutes of the ICAR are engaged in horticulture research. In addition, a few R&D establishments of crop/commodity boards and private sectors are providing research support to Indian horticulture. Research system in horticulture is now geared to provide necessary technological support to the expanding horticultural industry.

The research efforts in the past were mainly concentrated on crop improvement, propagation of seed/planting material, agrotechniques, crop protection and post harvest management. Some of the improved technologies developed are enumerated below.

Technology Generation

Varietal Development

Among the fruit crops, improved high yielding mango varieties, Mallika, Amrapali, Ratna, Sindhu, Arka Aruna, Arka Puneet, Dashehari-51 and hybrids namely CISH-M-1 and CISH-M-2 have been developed. Mallika is coming up in southern states like Karnataka and Amrapali is performing well in eastern India. Dashehari-51, a regular bearing cultivar with about 38 % higher productivity than the normal Dashehari, has been identified after 14 years of rigorous selection. In guava, three selections, namely Lalit, CISH-G-1 and CISH-G-2 have been developed for domestic and export markets. The fruits of Lalit are of medium size weighing about 150 g each and suitable for both table and processing purposes. In banana, high yielding hybrids like FHIA-01 and FHIA-03 are promising for replacing varieties, Panchananda and Bluggoe, respectively. Cultivar, Saba is found promising under sodic soils. In addition, high yielding varieties like Col, Hl and H2 have been developed. In grape, superior and high yielding varieties have been developed *e.g.* Beauty Seedless and Pusa Seedless and early ripening variety, Perlette for cultivation under North Indian conditions and Anab-e-Shahi, Dilkhush, Thompson Seedless, Tas-A-Ganesh, Sonaka, Bangalore Blue and Pachadraksha for south Indian conditions. In citrus, high yielding and cluster bearing varieties of acid lime have been developed. *e.g.*, Rough Lemon,

Rangpur Lime, Pramaini, Vikram PKM-1. Trifoliate oranges, namely Flying Dragon and Rich 16-6 are dwarfing types. In papaya, high yielding superior varieties both for table purpose and papain production have been developed. *e.g.*, Co-1 to Co-7, Coorg Honey Dew, Pusa Delicious, Pusa Majesty, Pusa Giant and Pusa Nanha. In apple, superior hybrids have been developed. *e.g.*, Lal Ambari, Sunehari. Red Spur, Star Crimson, Golden Spur, Red Chief, Oregon Spur, Skyline Supreme and Vance Delicious have been identified. Tissue culture protocols for micro-propagation of two commercial varieties have been developed.

Among the vegetable crops, more than 130 open pollinated varieties, 36 hybrids, 3 synthetics and 29 resistant varieties of 20 vegetable crops have been developed and released for cultivation in different agro-climatic regions. These include 40 in tomato, 45 in brinjal, 13 in cauliflower, 12 in chillies, 20 in pea, 9 in musk melon, 16 in onion and 44 in other crops.

In potato, 33 high yielding varieties have been developed indigenously for large scale cultivation in different regions. Kufri Ashoka and Kufri Pukhraj mature in 75 days. Kufri Jawahar and Kufri Satluj are field resistant to late blight. Kufri Jawahar has most ideal plant type for inter-cropping. Kufri Chipsona-1 and Kufri Chipsona-2 have been developed with excellent processing attributes, comparable to exotic varieties. Kufri Swarna resistant to golden nematode is ideal for Nilgiri Hills.

In tuber crops, improved varieties of different tuber crops have been recommended/released for commercial cultivation. These includes 9 varieties of cassava, 15 varieties of sweet potato, 6 varieties of colocasia, 3 varieties each of greater yam and lesser yam, 1 each of *Amorphophallus,* taro and yambean. Cassava varieties, Sree Visakham and Sree Prakash are popular in Kerala. Triploid clone, Sree Harsha with high dry matter and starch content is suitable for industrial belt of Tamil Nadu. Two early maturing varieties Sree Jaya and Sree Vijya have been released for culinary purposes. Elephant foot yam variety, Am-15 has been released with high yield potential of 41 t/ha. Among the plantation and spice crops, India is the first country to exploit hybrid vigour in coconut. Twelve hybrids involving tall and dwarf parents and 4 varieties have been released for commercial cultivation. These varieties yield 21 to 89 % more than the local cultivars. Some of the released varieties like Chandra Kalpa and Pratap (Banawali Green Round) are receiving wide acceptance of farmers. Chowghat Green Dwarf variety is good for tender coconut purpose. In arecanut, 4 high yielding varieties, namely, Mangala, Sumangala, Sreemangala and Mohitnagar have been developed, giving about 30 % higher yield than the local cultivars.

In oil palm, first efforts for improvement were made by producing Tenera hybrids using Pisifera pollen imported from Nigeria. Dura x Pisifera hybrids are field tested in East and West Godavari districts, Khammam and Krishna districts of Andhra Pradesh, with yield potential of 20-25 tonnes/ha of FFB from the fifth year. In cashew, 22 region specific selections and 12 hybrids with yield potential of 1.5-2 tonnes of raw nuts/ha have been produced and released for commercial cultivation. The present standards fixed for cashew varieties include export grade kernels of W-210 to W-240 and at least one tonne per ha yield with 30 per cent shelling. In black pepper, 6 varieties, namely, Sreekara, Subhakara, Palode-2, Panniyur-2, Panniyur-4, Panchami and Pournami, and 2 hybrids *viz.* Panniyur-1 and Panniyur-4 have been developed. In cardamom, a number of improved varieties have been developed and released for commercial cultivation *e.g.,* CCS-1, Mudigere-1, PV-1, ICRI-1 and ICRI-2. In ginger, varieties like Suprabha, Suruchi, Suravi and Varada have been developed. In turmeric, several varieties *viz.,* Co-1, Krishna, Sugandham, BSR-1, Suvarna, Roma, Suroma, Rajendra Sonia, Sugana, Sudarshana, Ranga, Rasmi, Prabha, Prathiba, Mega Turmeric and RCT-1 have been developed with yield potential of up to 44 tonnes of fresh rhizomes per ha. Three high yielding cinnamon lines, namely, Navashree, Nithyashree and Konkan Tej have been released for cultivation.

Seed/planting Material Propagation

In most of the fruit crops, vegetative propagation techniques have been standardized. Soft wood grafting has been standardized for mango, sapota, custard-apple and jackfruit. Other vegetative propagation techniques have been developed for *ber, aonla,* jackfruit, custard-apple and *bael.* In mango, veneer grafting and stone-grafting is practised commercially. Mango variety, Vellaikolumban is suitable semi-dwarfing rootstock for Alphonso. Old unproductive mango trees can be rejuvenated successfully by pruning the 4th order branches during December-January. Flowering and fruiting are regular in pruned trees. For mandarin orange, Rangpur lime is a drought hardy rootstock. In grapes, Dogridge and Salt Creek (Ramsey) are suitable for minimizing adverse effects of soil salinity on Thompson Seedless. A tissue culture technique for mass multiplication of Dogridge has been standardised. In banana, sword suckers of 700-1000 g are optimum. Rhizomes with active lateral buds and dead central buds are preferred for distant transportation in western India. Double paring and shade drying followed by dipping in Monocrotophos (0.5 %) and Bavistin (0.2 %) is recommended to disinfect nematodes and soil borne fungi. Tissue cultured banana plants are now commercially adopted for their uniformity in flowering and produce. Shoot-tip grafting technique in citrus has been considerably advanced.

In vegetable crops, seed production of over 120 open pollinated high yielding varieties of different vegetables has been well established in the country. Hybrid seed production has become easier with the development of male sterile lines in tomato, self incompatible lines in cauliflower and gynoecious lines in cucumber and muskmelon. In brinjal, functional male sterility controlled by a single recessive gene has been identified. Temperature barrier in cole crops (cabbage and cauliflower) has been overcome by developing heat tolerant hybrids. It is now possible to cultivate cabbage and cauliflower in southern India. Development of tomato varieties resistant to bacterial wilt has made their cultivation successful in non-traditional areas. Onion seed production technology for cultivation in kharif season has been developed for north Indian states especially, Haryana, Punjab and western Uttar Pradesh. Seed Plot Technique has been developed for production of disease-free potato seed in plains. It is widely adopted by farmers. A new technology for raising commercial crop of potato using 'True

Potato Seed' (TPS) has been developed and standardized as supplementary technology to the traditional tuber grown crop. Two TPS populations, TPS-C-3 and HPS-113 are recommended for commercial production in Bihar, Gujarat, Tripura and West Bengal. Micropropagation protocols have been developed in banana, oil palm, cashew, black pepper, ginger, *etc*. Seed gardens of Tall (T) x Dwarf (D) and D x T hybrids have been established for production of coconut hybrids.

Agrotechniques

In fruit crops, improved agrotechniques developed have helped the farmers in improving the productivity and quality of produce. Soil application of paclobutrazol (4 g/tree) increase flowering and fruiting in mango on commercial scale in coastal Maharashtra. It also controls irregular bearing in cultivar Dashehari. Spray of NAA @ 200 ppm in October is recommended for control of malformation. Heavy fruit drop at maturity in cultivar Langra can be controlled by spraying NAA (20 ppm). In guava, double spray of 10 and 20 % urea on cultivars, Allahabad Safeda and Sardar twice at bloom eliminate poor quality rainy season crop and increases winter season yield by 3 and 4 times, respectively. Application of neem coated urea (800 g/plants) yields 98 kg fruits/plant in variety, Sardar compared to 37 kg from untreated plants. In banana, high density planting (4550 plants/ha) yield up to 174 t/ha. Adoption of improved technology in Maharashtra has resulted in fruit yield increase up to 52 t/ha. In citrus, two grafting methods using inverted 'T' cut and apical triangle cut have been developed with overall success of around 36% using either method. For accelerating the survival of growth, shoot tip grafts, the successful grafts are double grafted (side grafted) on vigorous greenhouse grown Rough Lemon and Rangpur Lime seedlings. Rangpur Lime rootstock is found superior for sweet oranges and mandarins. In papaya, closer spacing of 1.4 x 1.4 m is recommended for high yield. Drip irrigation techniques have been standardized. In banana, it has resulted in production gain (60-70%) and early harvesting (40-50 days), besides improved water efficiency. Likewise, in grapes higher yields have been obtained with better water use efficiency (11 %).

In vegetable crops, improved production technology has been developed for major crops. Drip irrigation is economical in tomato and brinjal. In cucumber, replenishment of evaporation loss through irrigation resulted in maximization of yield of quality fruits. Drip irrigation in watermelon provided 33% higher yield with a water saving of 40%. Nutrient requirements and fertiliser schedules have been worked out crop-wise and recommended for different agro-climatic regions. In leguminous vegetables, high N depresses nodulation. The VAM fungi increases P availability to plants. In all leguminous vegetables, inoculation of the VAM fungi along with *Rhizobium* culture is beneficial. Production technologies for *kharif* onion in northern India and long day type onions for high altitudes have been standardized. Pendimethalin (Stomp) has been found effective in controlling weeds in tomato, brinjal, chilli, bell pepper and okra. In potato, a number of potato-based multiple and inter cropping systems have been developed for different potato growing regions. Intercrop combinations with sugarcane in Maharashtra, wheat in Chhota Nagpur area and linseed in central Uttar Pradesh are found remunerative. A suitable method of urea application has also been worked out.

In tuber crops, short duration legumes *viz.*, groundnut and French bean and cowpea can be successfully inter-cropped with cassava. Short-duration cassava, Sree Prakash is ideal in double cropped rice fields. Studies on cassava-based multiple cropping systems involving banana, coconut, *Leucaena* and Eucalyptus, have shown banana-cassava combination to give maximum root yield. Banana and coconut combination reduces soil loss and surface run-off considerably. *Dioscorea* and elephant foot yam with banana, Nendran can generate an additional income of Rs 20,000/- over the sole crop of banana. For Inter-cropping *Dioscorea* in coconut garden, the ideal planting density is 9000 plants/ha. When *Amorphophallus* is raised as an inter-crop in coconut garden, one third dose of recommended fertilizers is sufficient. Inoculation with VAM fungi in cassava give about 15-20% increase in yield.

Among the plantation and spice crops, density of 175 coconut palms/ha (7.5 m x 7.5 m spacing) is found ideal. In general, NPK application of 500:320:1200 g/palm/year is found optimum.

A multi-storied cropping system involving black pepper trained on coconut trees, and cocoa in between the rows of coconut and pineapple in the ground floor has been found ideal for exploiting light, soil and air spaces. In arecanut, application of NPK (100:40:140 g) and green leaf (14 kg) per palm per year is recommended for coastal regions of Kerala and Karnataka and for plains of West Bengal, Karnataka and Assam. In oil palm, application of NPK (1200:600:1200 g/palm) is found to give 17.1 tonne of FFB per ha. In black pepper, rapid methods for production of rooted cuttings of pepper have been developed and a commercial protocol has been standardised for micropropagation of black pepper. Ginger yield could be increased up to 33 percent by application of neem cake at the rate of two tonnes per ha and the fertiliser schedule of 75 kg each of N, $P_2 0_5$ and $K_2 0$. Technology for storage of ginger seed rhizome is standardised and recommended.

Protection Technologies

For major fruit crops, plant protection schedules have been developed for the control of significant insect-pests for wider adoption. In the recent years, research efforts are directed to devise eco-friendly, economical and long lasting control measures. Success has been achieved in biological control of mealy bugs in mango and guava. The *Beauveria bassiana* has been found killing mango mealy bug and hopper. In grapes, integrated management of *Spodoptera* caterpillar involving light and pheromone traps, NPV and neem based insecticides and biological control of mealy bug by the beetle *Cryptolaemus montrouzieri* have been standardized. Studies on pesticide residues have resulted in working out of safe-waiting periods for harvesting and consumption of fruits.

In vegetable crops, about 50 improved measures for efficient management of diseases and 23 for insect-pests have been worked out and popularized in different agro-climatic regions in the country. Integrated pest management (IPM) for controlling diamond back moth on cabbage through a trap crop like mustard has been demonstrated. Fruit borer *(H. armigera)* on tomato can be controlled by the release of *Trichogramma pretiosum* alone and in combination with HaNPV. In potato, integrated management schedules for control of bacterial wilt and tuber moth have been developed. A late blight forecasting system has been developed for the hills.

Among the tuber crops, the major diseases affecting tuber crops are cassava mosaic and brown leaf spot in cassava, *Phytophthora* leaf blight in colocasia, *Fusarium* wilt in elephant yam and virus diseases of sweet potato. Foliar sprays of Bavistin (0.1%) combined with disodium and dipotassium phosphates (100 ppm) and calcium sulphate at 15-day interval was found to check foliar diseases in sweet potato. The major pests include spider mites, scale insects and white fly on cassava, weevil on sweet potato, defoliators, aphids and mites on colocasia, and scales and mealy bugs on yams and elephant yam. Cultural methods for weevil control include clean cultivation, destruction of alternate hosts and timely harvest. An effective IPM package using synthetic sex pheromone has been developed. Control measures involving insecticides have been evolved for the control of pests of other tuber crops.

Among the plantation and spice crops, bud rot of coconut caused by *Phytophthora palmivora* can be effectively controlled by spraying Bordeaux mixture. Calyxin root feeding and drenching of soil with 1% Bordeaux mixture along with neem cake application @ 5 kg per palm per year is recommended for controlling Thanjavur wilt disease reported in Tamil Nadu, Andhra Pradesh and Karnataka.

A package of practices has been developed for managing mycoplasma like organisms (MLOs) in root wilt affected coconut palms in Kerala and Thatipaka disease affected palms in Andhra Pradesh. Eradication of all root wilt affected palms is recommended. In cashew, tea mosquito bug (TMB) can be effectively controlled through a schedule of spray coinciding with flushing, flowering and fruiting. For effective control of stem and root borer infestation, constant monitoring and adoption of strict sanitation in the plantations coupled with prophylactic application of coal tar and kerosene in the ratio of 1:2 on trunks are recommended. In black pepper, spraying Bordeaux mixture (1 %) and drenching the soil with copper oxychloride (0.2 %) is found effective in managing *Phytophthora* foot rot.

Post Harvest Management Technologies

The post harvest handling of fruits and vegetables accounts for 20-30% of losses at different stages of storage, grading, packing, transport and finally at marketing as a fresh produce or in processed form. A number of improved technologies have been developed for commercial exploitation. An on-farm, low cost, environment friendly cool chamber, Zero Energy Cool Chamber has been developed using locally available material. The principle of evaporative cooling reduces the inside temperature by as much as 17-18 °C and keeps the relative humidity above 90% during peak summer. The chamber increases the shelf life and reduces PLW of banana, mango, orange lime, grape fruit, tomato and potato in different situations in India. Maturity standards for mango, guava, grape, litchi and *ber* and chemical treatments for regulation of ripening in mango, sapota and banana have been standardised. Optimum storage temperatures worked out for several fruits, vegetables and tuber crops. A mango harvester, fruit peeler, hand and pedal operator cassava chipping machines, harvesting tools (5-14 times efficient), coconut dehusking machine, implements for mechanization of potato cultivation and other crops have been developed. A number of improved technologies have been developed for commercial exploitation *viz.*, tent type foldable solar dryer, packaging boxes for distant transportation of apple, mango, citrus and plum, production of value added products-pectin from peel and flour from mango fruit kernel, production of fruit post carbonated beverages *etc.*

Production Constraints

In spite of great strides made, the productivity of horticultural crops, in general, is still quite low and the post harvest losses particularly of perishable commodities, are considerable. Improvement in quality standards of the produce and their marketing are essential to increase our share in the global market. The research agenda in horticulture is by design relevant to national plans and priorities and research programmes are normally formulated keeping in view the thrust areas in development. The major technology related constraints contributing to low productivity of horticultural crops and inferior quality of produce are:

- Vast majority of holdings are small and un-irrigated.
- Large tracts of low and unproductive plantations needing replacement/rejuvenation.
- Low productivity of crops due to inferior genetic stocks and poor management.
- Inadequate supply of quality planting materials of improved varieties.
- High incidence of pests and diseases.
- Heavy post harvest losses and low utilization in processing sector.

For addressing the above constraints, research institutions are engaged in both basic and applied research. While formulating research strategies some of the inherent weaknesses associated with perennial tree crops and certain perpetual problems in Indian horticulture must be kept in mind. They are:

- Long period required for development of improved genotypes. Application of biotechnological tools/methods in horticultural crops is still in its early stage of development in the country.
- Chronic production problems due to major disorders like alternate bearing, malformation and spongy tissue in mango, guava wilt, citrus decline, root wilt in coconut, viral disease in vegetables, *Phytophthora* diseases in large number of crops *etc.* remained largely unresolved.
- Lack of advanced technologies for post harvest handling, processing and marketing of produce.

Losses caused by biotic stresses are very high and due to pesticide residue problems development of eco-friendly IPM strategy is more relevant in horticulture. There is a threat for loss of valuable genetic resources, if measures are not taken for their conservation. Wastelands and hilly terrains being the potential future expansion areas, matching technologies for dry land and hill horticulture need to be developed. Counter seasonal advantages from diverse agro-climatic situations provide strength for extended availability of horticultural crops round the year and such potentials can be harnessed only with relevant research support.

2

Research Strategies and Programmes

Keeping in view of the strength and weaknesses of the research system and priority areas of horticulture development the following strategies and programmes are suggested for horticultural research:

Rationalization of Research

There is a need to shift from commodity/discipline oriented research to system based research and to establish stronger Inter-institutional linkages with SAUs, CSIR, DBT and BARC. Greater private sector partnership for diversification, value addition and export promotional research and seed production programmes will be required to modernize horticulture industry. Introduction of project based budgeting in the Institutes/NRCs will bring better accountability and one time catch up grant to modernize old Institutes will be essential. Development of database on technologies evolved, market intelligence, export projections/removal of quantitative restrictions, R&D scenario in horticulture should receive priority attention.

Safeguard for Intellectual Property Rights (IPR)

A fool proof description of varieties and their registration and finalisation of material transfer agreement and channelization of germplasm exchange need to be institutionalized. Similarly, specification of quality/codex standard for export of indigenous fruits and vegetables need to be developed. Phytosanitary regulations for importing vegetatively propagated materials need a relook/revision and rigorous enforcement.

Genetic Resource Management

Greater emphasis need to be given on (a) *in situ* conservation of endangered species, *ex situ* conservation of base collections and *in vitro* storage and cryopreservation of important germplasm. Clonal repository of vegetatively propagated crops and germplasm screening for processing and diversified use will be certain other areas of priority research.

Horticultural Biotechnology

Biotechnology as a tool for rapid multiplication of quality planting material, virus cleaning, genetic transformation *etc.*, will be of very great importance. The priority research programmes are:

- Development of micropropagation protocols in selected crops
- Genetic engineering for integration of desirable traits
- Molecular characterization of germplasm and development of molecular linkage maps
- Value addition to products
- Preservation of post harvest losses through control of metabolic process

Integrated Production System

Production related technologies can bring quick improvement in production and productivity on different regions. Short or medium range programme on horticulture based cropping systems; water management including microirrigation and fertigation, greenhouse cultivation of vegetables and flowers, integrated nutrient and pest managements, environment pollution and pesticide residue problems have already received research attention. Further refinement of the technologies and their transfer will bring perceptible change/improvement in production of different crop commodities.

Export Promotional Research

There is a need for development of bulk handling system of tropical fruits, including pre-cooling and CA/MA storage and post harvest protocols for sea transport of major fruits like banana, mango, litchi, sapota, Kinnow and pomegranate. Disinfestation technology including vapour heat treatment (VHT) for export of fresh fruits and extension of shelf life by preventing desiccation of vegetables should help in further export promotion. Organic farming for vegetable and spice crops and residue free IPM technology are other important areas of research.

Quality Planting Material Production

For resolving the long standing problems of supply of good quality planting materials of different horticultural crops, following research areas are flagged:

- System of certification and standardisation of planting materials;
- Micropropagation protocols for mango, litchi, coconut, walnut, date palm, oil palm and apple rootstocks;
- Refinement of TPS and micro-tuber production technology in potato;
- Improvement in STG techniques and cross-protection in citrus for virus elimination and control; and
- Seed production system for hybrid vegetables and commercialization of micropropagation in floriculture.

Post Harvest Management

In order to reduce post harvest losses at production centres low cost eco-friendly on-farm storage structures can play a crucial role. Significant advancement has been made in that direction and some small and medium sized cool chambers on the principles of evaporative cooling have been devised. Further refinement of the technology will go on a long way. Also, standardisation of packing line operations and proper packaging of different commodities are of urgent need. Pesticide residue management and newer product developments will add values to the produce.

Human Resource Development

Advanced training in research methodologies and instrumentation, biotechnology, micro-irrigation, fertigation,

IPM, INM, biofertilizer, biopesticide, pesticide residue, PHT and product development need priority attention for increasing research capabilities of the scientists. Skill development for state level development functionaries through in-service training at different R&D institutions will enhance capabilities of extension staff. Post Graduate programmes in fruit, vegetable, floriculture, plantation crops and post harvest management of horticultural crops will help in providing trained manpower in specialized areas.

Technology adoption pattern can not be uniform throughout the country and will vary from crop to crop and even from region to region. Certain degree of flexibility in research planning and research strategy is therefore obvious. Also, with the opening of global markets and removal of quantitative restrictions under the WTO export-import scenario is likely to change at much faster pace.

Market forces will play a more dominant role and demands for modern technologies will increase. Research system in horticulture will have to be very alert and should be able to adjust with the changes. Development of both short term and long term strategies for modernising Indian horticulture will depend largely on the research support and strength of research system.

Neglected Horticultural Crops

The processes and causes of the marginalization of Iberian crops, more than 20 horticultural crops are mentioned which could be considered to be in this situation. The authors have selected eight which will be dealt with in detail. Selection was based on a stricter identification of their marginalized nature and choosing from various taxonomic groups that would allow a detailed view of the problem.

Rocket (*Eruca sativa*), garden cress (*Lepidium sativum*), purslane (*Portulaca oleracea*), borage (*Borago officinalis*), alexanders (*Smyrnium olusatrum*), scorzonera or black salsify (*Scorzonera hispanica*), spotted golden thistle (*Scolymus maculatus*) and Spanish salsify or Spanish oyster plant (*Scolymus hispanicus*) are the eight species selected.

Rocket (Eruca saliva)

Botanical Name: *Eruca sativa* Miller

Family: Brassicaceae = Cruciferae

Common Names: English: rocket, salad rocket, garden rocket; Spanish: oruga, oruga común, eruca, roqueta común; Catalan: ruqueta; Basque: bekarki; Portuguese: eruca, rúcula, fedorenta, pinchão (Brazil); French: roquette

Origin of the Name: The semantic origin of this plant's name alludes to the oldest crops of the Near East. The Persian *girgir* and Acadian *gingiru* gave the Aramaic, Hebrew and Syrian *gargira*, and from these the Arabic *yiryir* and Latin *eruca*, from which, through Spanish, the words "roqueta" and "oruga" of present-day Spanish appeared.

Properties and Uses: This plant is considered to be an excellent stomachic, stimulant and aphrodisiac, and is also used as a diuretic and antiscorbutic. The leaves have a bitter flavour which is made milder by cooking or frying. The seeds are hot, although rather less so than mustard seeds. It contains glucosides, such as allyl sulphocyanate, mineral salts and vitamin C. The oil of the seed contains erucic acid.

Rocket was always considered to be a potent aphrodisiac. In classic antiquity, it was consecrated to Priapus and was planted at the foot of the statue of this deity dedicated to the procreative potential of males. Dioscorides warns that, eaten raw, it stimulates lust and that the seeds have the same power. Columela also refers to its sexually stimulating effect, but is also very well acquainted with its cultivation technique: "...and rocket and basil also remain in the place where they have been sown and require no other care than manuring and weeding. Moreover, they can be sown not only in autumn, but also in spring...." The Hispano-Romans also compared the aphrodisiac power of rocket precisely with the anaphrodisiac power of lettuce. In Hispano-Visigoth culture, Isidoro de Sevilla supports the use and knowledge of this plant's powers: "... rocket is, so to speak, inflammatory, since it has burning properties and, if consumed frequently in the diet, arouses the sexual appetite. There are two species, one of which is in habitual use while the other is wild with a more bitter taste. Both stimulate sexual appetite."

Irrespective of these effects, rocket has been eaten basically as a vegetable (leaves) and as a spice (leaves and seeds). It is thus an ingredient of misticanza" (mixed salad), a speciality eaten in Rome since the very foundation of that city. Hispano-Arab agronomists also mention its cultivation, for instance Ibn Hayyay (eleventh century). Ibn Wafid (eleventh and twelfth century) and of course, Ibn al-Awwam (twelfth century). The latter author mentions the plant's use as a flavouring for musts and syrups, the seed being ground and scattered over the surface of the earthenware jars containing the syrup. He also mentions its flowers being used in a similar way. In the sixteenth century, Alonso de Herrera's *Tratado de agricultura* contains no mention of rocket. It is used to make sauces in which the leaves are mixed with sugar or honey, vinegar and toasted bread (rocket sauce). In Italy, it is eaten boiled with spaghetti, and then seasoned with garlic and oil. In Spain, it is traditionally used in La Roda and Montealegre del Castillo (Al bacete) in the preparation of *gazpachos* of La Mancha, an ancestral dish which includes the meat of partridge and rabbit and unleavened bread (*gazpacho*), with lightly fried rocket. Some authors relate this tradition to primitive fertility rituals.

Nowadays it still remains very much appreciated in various countries of the Mediterranean area including Italy, Greece and Turkey, where it is eaten mainly in salads and as a garnish for meat. It goes very well with lettuce, chicory, valerian and tomato. Another recipe is potato and rocket salad. In India, it is cultivated to obtain a semi-drying oil from the seeds. At present, most of the rocket grown is for this purpose, and it is considered mainly as a potential oilseed product.

This plant's marginalization as a vegetable in Spain may have been very much connected with its condemnation because of its aphrodisiac properties.

Botanical Description

Rocket is an annual herbaceous plant, growing up to 80 cm. The basal leaves occur in a rosette and are lyrate-pinnatifid (those normally eaten in salads); the caulinar leaves are lobulate or dentate. The flowers have white or light yellow petals. The siliquae measure up to 40 mm, are erect, attached to the stem, with a subcylindrical valvar portion and an ensiform face as long as the valves. The seeds measure 1.5 to 2.5 mm and are brown. The wild form flowers from February to June and the cultivated form right into mid-summer. It is allogamous with a complex system of self-incompatibility, mainly gametophytic, but with some alleles acting sporophytically. The existence of genie male sterility has been verified. The chromosome pattern is 2n = 2x = 22.

Horticultural crops: A) rocket (Eruca saliva), detail of fruit in the silicle; B) garden cress (Lepidium sativum), detail of fruit in the silicle; C) purslane (Portulaca oleracea)

Ecology and Phytogeography

Rocket grows spontaneously in places modified by humans: abandoned gardens, waysides, tips and among rubble. It prefers hot, dry climates. It is distributed all around the Mediterranean, extending to central Europe in the north and as far as Afghanistan and northern India in the east. It has reverted to the wild state in North America, South Africa and Australia. Vavilov described it in central Asia, the Near East and the Mediterranean, the latter being considered its main centre of origin. It is cultivated mainly in India, and is grown more rarely in Turkey and Greece. It is also cultivated in Italy. In countries such as Spain, France and Great Britain, cultivation is rare.

Genetic Diversity

The biggest collections of rocket germplasm are to be found at the Institute of Germplasm in Bari, Italy, at the NBPGR in New Delhi, India, at the Haryana Agricultural University in India and at the VIR in St. Petersburg.

There are also smaller collections in Kabul in Afghanistan, Saskatoon in Canada, Gaersleben and Braunschweig in Germany, Tapioszele in Hungary, Islamabad in Pakistan, Blonie in Poland and Alnarp in Sweden. A small collection of species of *Eruca*, including *E. sativa*, is to be found at the Universidad Politécnica de Madrid and there is also germplasm from wild populations of the genus at the Córdoba Botanical Garden.

Collecting expeditions have continued. In 1985, 25 samples of indigenous germplasm of *E. sativa* were collected in the northeastern Sudan.

In an analysis using the D^2 statistic of Mahalanobis, out of 99 lines of rocket no correlation was found between genetic diversity for 12 characters associated with production and geographical origin.

There is wide variability as regards the characters of the siliqua and its stability, and a strong interaction with the cultivation conditions. Similarly, there is wide genetic variability for seed production per plant and related characters.

An important group of studies is attempting to use *E. sativa* as a genetic resource for improving other crucifers. In this way, intergeneric hybrids have been obtained with *Raphanus sativus*, *Brassica campestris* and *B. oleracea*. Somatic hybrids have been obtained through the fusion of protoplasts with *B. napus* and *B. juncea*.

There are lines of rocket (T27) known which are resistant to mustard aphid and tolerant of several stress conditions as well as *Fusarium oxysporum*. Such lines may also be a source of genes that are transferable to species of *Brassica*.

Cultivation Practices

Rocket is a very hardy plant which requires little care. It is generally sown direct in late winter or early spring, in shallow furrows. To encourage emergence, it is advisable to cover with light sieved soil. It requires little irrigation and manuring. It is usually hoed by hand.

The young leaves are harvested in spring.

Prospects for Improvement

The use of rocket as a vegetable, salad or spice has been marginalized, possibly for moral or religious reasons, and its recovery is limited by local gastronomic tradition, which is not always able to appreciate its characteristic bitter flavour. This is due to glucosinolates and the high content of mineral salts.

The development of cultivars with a low allyl sulphocyanate content does not appear to be an improvement objective since, even though the plant would be rendered innocuous, it would lose its individual identity.

In fact, a wide variability has been observed as regards both erucic acid content and glucosinolate content in 128 rocket specimens from Pakistan. Rocket already has a low content of these constituents, and the local inhabitants clearly distinguish this species from other more bitter crucifers. Its use can be increased only through the promotion of the traditional dishes in which it appears.

The use of agronomic techniques such as nitrogen fertilization and shading would enable younger, more juicy rosettes to be obtained which have a milder taste and are more palatable.

The work on genetic improvement for the use of "rocket" as a vegetable is very limited, if we exclude the development of in vitro cultivation, which has made it possible to regenerate normal diploid plants from isolated protoplasts of leaf mesophyll.

Garden Cress (Lepidium sativum)

Botanical Name: *Lepidium sativum* L.

Family: Brassicaceae = Cruciferae

Common Names: English: cress, common cress, garden cress, land cress, pepper cress; Spanish: mastuerzo, mastuerzo hortense, lepidio, berro de jardín (Spain), berro de tierra, berro hortense (Argentina), escobilla (Costa Rica); Catalan: morritort, morrisá, Portuguese and Galician: masturco, mastruco, agrião-mouro, herba do esforzo; Portuguese: mastruco do Sul, agrião (Brazil); Basque: buminka, beatzecrexu

Origin of the Name: Cultivation of this species, which is native to Southwest Asia (perhaps Persia) and which spread many centuries ago to western Europe, is very old, as is shown by the philological trace of its names in different Indo-European languages. These include the Persian word *turehtezuk*, the Greek *kardamon*, the Latin nasturtium and Arabic *tuffa'* and *hurf*. In some languages there is a degree of confusion with watercress.

It seems that the meaning of the word nasturtium (*nasum torcere*, because its smell causes the nose to turn up) must have been applied initially to garden cress, as both Pliny and Isidoro de Sevilla explain. The confusion remains with the terms used by the Hispano-Arabs. The word *hurf* is applied without distinction to watercress and garden cress (several species certainly of up to three different genera: *Nasturtium*, *Lepidium* and *Cardaria*). Thus the medieval agronomists of Andalusia went as far as differentiating between several *hurf*, such as *hurf abyod, hurf babili, hurf madani*....

Properties, Uses and Cultivation

Xenophon (400 BC) mentions that the Persians used to eat this plant even before bread was known. It was also familiar to the Egyptians and was very much appreciated by the Greeks and Romans, who were very fond of banquets rich in spices and spicy salads. Columela (first century) makes direct reference to the cultivation of garden cress. In *Los doce libros de Agricultura*, he writes: "...immediately after the calends of January, garden cress is sown out... when you have transplanted it before the calends of March, you will be able to harvest it like chives, but less often... it must not be cut after the calends of November because it dies from frosts, but can resist for two years if it is hoed and manured carefully... there are also many sites where it lives for up to ten years" (Book XI). The latter statements seem to indicate that he is also speaking of the perennial species *L. latifolium*, as *L. sativum* is an annual.

Almost all of the Andalusian agronomists of the Middle Ages (Ibn Hayyay, Ibn Wafid, Ibn alBaytar, Ibn Luyun, Ibn al-Awwam) and many of the doctors, such as Maimonides, mention garden cress. Ibn al-Awwam also includes references from Abu al-Jair, Abu Abdalah as well as from Nabataean agriculture and, among other comments, he says: "Garden cress is sown between February and April (in January in Seville). It has small seeds which are mixed with earth for sowing to prevent the wind carrying them away.... It is harvested in May and is grown between ridges, in combination/conjunction with flax cultivation."

Many of the authors of the old oriental and Mediterranean cultures emphasized the medicinal properties of cress, especially as an antiscorbutic, depurative and stimulant. Columela notes its vermifugal powers. Ibn al-Awwam refers to certain apparently antihistaminic properties, since it was used against insect bites and also as an insect repellent, in the form of a fumigant. It was perhaps Ibn al-Baytar, an Andalusian botanist (eighth century), who collected most information on its properties, summarizing the opinions of other authors such as El Farcy, who says that it incites coitus and stimulates the appetite; Ibn Massa, according to whom it dissipates colic and gets rid of tapeworms and other intestinal worms: or Ibn Massouih, who mentions that it eliminates viscous humours. Ibn al-Baytar also says that it is administered against leprosy, is useful for renal "cooling" and that, if hair is washed with garden cress water, it is "purified" and any loss is arrested.

In Iran and Morocco, the seeds are used as an aphrodisiac. In former Abyssinia, an edible oil was obtained from the seeds. In Eritrea, it was used as a dyestuff plant. Some Arab scholars have attributed garden cress's reputation among Muslims to the fact that it was directly recommended by the Prophet.

Garden cress's main use was always as an aromatic and slightly pungent plant. Not only in antiquity but also in the Middle Ages it enjoyed considerable prestige on royal tables. The young leaves were used for salads. The ancient Spartans ate them with bread. This use still continues and they are also eaten with bread and butter or with bread to which lemon, vinegar or sugar is added. However, it is mainly used nowadays in the seedling stage, the succulent hypocotyls being added to salads and as a garnish and decoration for dishes. The roots, seeds and leaves have been used as a spicy condiment. Columela explains how *oxygala*, a type of curd cheese with herbs, was prepared: 'Some people, after collecting cultivated or even wild garden cress, dry it in the shade and then after removing the stem, add its leaves to brine, squeezing them and placing them in milk without any other seasoning, and adding the amount of salt they consider sufficient.... Others mix fresh leaves of cultivated cress with sweetened milk in a pot.

L latifolium L. stands out for its horticultural interest; although it grows spontaneously on the edges of rivers and lakes, it is also occasionally grown in the same way as *L. sativum*. Its young leaves can be used for salads; the ancient Greeks and Romans used to grow it for this purpose. Its leaves and seeds were also used as a spicy condiment. Several sauces are prepared with its leaves, including in particular the bitter sauce of the paschal lamb of the Jews. The seeds of this species were known in England as the poor people's pepper. The roots have been used on occasion as a substitute for radish.

In the fifteenth century, we know through Alonso de Herrera that garden cress was one of the vegetables most widely eaten in Castile. During the sixteenth century, obstinate attempts were made to introduce it into America. Right up to the beginning of the nineteenth century, its cultivation in Spain continued to be important, since Boutelou and Boutelou (1801) deal specifically with this crop in their *Tratado de la huerta*, commenting on the existence of several cultivars At present, the cultivation of cress is very occasional in countries such as Spain and France. Water cress, in competition with garden cress, has eclipsed the cultivation of the latter. However, this is not the case in other central European countries or the United Kingdom, where its use is normal and the system of cultivation has changed substantially.

Botanical Description

Cress is an annual, erect herbaceous plant, growing up to 50 cm. The basal leaves have long petioles and are lyratepinnatipartite; the caulinar leaves are laciniate-pinnate while the upper leaves are entire. The inflorescences are in dense racemes. The flowers have white or slightly pink petals, measuring 2 mm. The siliquae measure 5 to 6 x 4 mm, are elliptical, elate from the upper half, and glabrous. Cress flowers in the wild state between March and June.

It is an allogamous plant with self-compatible and self-incompatible forms and with various degrees of tolerance to prolonged autogamy. There are diploid forms, $2n = 2x = 16$, and tetraploid forms, $2n = 4x = 32$. A degree of variability is noted in the character of the basal leaves which are cleft or split to a greater or lesser degree, a character which is controlled by a single incompletely dominant gene.

Ecology and Phytogeography

Cress is a plant that is well suited to all soils and climates, although it does not tolerate frosts. In temperate conditions, it has a very rapid growth rate. It grows subspontaneously in areas transformed by humans, close to crops or human settlements. It appears in this way on the Iberian peninsula, mainly in the eastern regions.

Wild cress extends from the Sudan to the Himalayas. Most authors consider it to be a native of western Asia, whence it passed very quickly to Europe and the rest of Asia as a secondary crop, probably associated with cultivars of flax. Vavilov considers its main centre to be Ethiopia, where he found the widest variability: the Near East, central Asia and the Mediterranean are considered secondary centres. It is now naturalized in numerous parts of Europe, including the British Isles.

Genetic Diversity

The genus *Lepidium* is made up of about 150 species, distributed throughout almost all temperate and subtropical regions of the world. On the Iberian peninsula and the Balearic Islands, at least 20 species or subspecies exist among the autochthonous and allochthonous taxa, some genetically close to *L. sativum*. Seven of them are exclusively endemic to the peninsula or, at the very most, are common with North Africa. Other close species are *L. campestre* (L.) R. Br. and *L. ruderale* L.

which also have edible leaves. The leaves of *L. campestre* are used to prepare excellent sauces for fish.

Common cress (*L. sativum* L.), with regard to the anatomy of the leaf, stem and root, has been divided into three botanical varieties: *vulgare*, *crispum* and *latifolium*. The latter is the most mesomorphic, *crispum* the most xeromorphic and *vulgare* intermediate.

At present, most of the studies on the variability and development of new cultivars are being carried out in liaison with the VIR of St. Petersburg, where there is a good collection of material. Of the 350 forms of garden cress studied in the Ukraine, Uzkolistnyti 3 was the best, being highly productive and of good quality. It is being used as the basis of improvement programmes, as it appreciably surpasses the best Soviet varieties in production and quality. Other cultivars well suited to European Russia are Tuikers Grootbladige (broad-leaved) and the lines Mestnyi k137, k106 and k115. Of the types most cultivated in Europe, Early European, Eastern, Dagestan and Entire Leaved stand out, being distinguished by the length and shape of the leaf, earliness and susceptibility to cold. In Western Europe, one broad-leaved type is especially appreciated (Broad Leaved French) as are curly types (Curly Leaved), the latter being used extensively to garnish dishes. In Africa, there are red, white and black varieties.

This crop is also arousing interest in Japan, and collecting expeditions to Nepal have been organized. Some specimens collected during an expedition to Iraq in 1986 are now stored in Abu Ghraib and in Gratersleben, Germany. There are also small collections of *L. sativum* in the PGRC in Addis Ababa (Ethiopia), at the ARARI of Izmir in Turkey and in Bari, Italy. At the Universidad Politécnica de Madrid there are accessions of 20 species of *Lepidium*, while the BGV of the Córdoba Botanical Garden keeps germplasm of the southern Iberian species of the genus.

Cultivation Practices

Cress is an easily grown plant with few requirements. It can be broadcast after the winter frosts or throughout the year in temperate climates. However, Boutelou and Boutelou (1801) were already recommending sowing in shallow furrows, which enables surplus plants to be thinned out and facilitates hoeing. Sowing has to be repeated every 15 to 20 days so that there is no shortage of young shoots and new leaves for salads—the leaves of earlier sowings begin to get tough and are no longer usable. The seed sprouts four or six days after sowing, depending on the season, and the leaves are ready for consumption after two or three weeks.

The usual form of cultivation continues to be as described, with 15 to 20 cm between rows and the use of irrigation in the summer, since they are lightly rooted seedlings which can dry up in a few days. Its growth is very rapid and harvesting can begin in the same month as sowing, with yields reaching 6 tonnes per hectare.

Prospects for Improvement

Most of the genetic improvement work on garden cress is being carried out in the CIS, with little or no work being done at present in the countries of western Europe. Mainly early cultivars with a prolonged production period and better cold tolerance are being developed.

Cress can be grown and used like white mustard. It geminates more slowly at low temperatures, the emergence period being three or four days longer. Shortening this period is an interesting improvement objective.

However, cress's recovery and its greater presence on markets mainly depends on a modification of cultivation and marketing techniques. In counties such as the United Kingdom, where this vegetable is nominally to be found at the markets, cultivation takes place in greenhouses throughout the year. The whole succulent hypocotyls of the very young seedlings are eaten. The seed is placed on the soil surface on soft, level beds. It is finely sprinkled with water and then covered with sackcloth which has been steam-sterilized and moistened. The latter is frequently wetted to maintain moisture and is removed when the seedlings reach 4 to 5 cm in height (after approximately seven days in spring and autumn and ten days in winter). The yellowish leaves turn green after two to three days. The cress is harvested when the first pair of cotyledon leaves have developed and it is marketed in small bags or trays, sometimes together with seedlings of white mustard.

Garden cress and white pepper are sometimes sown in the plastic trays or bags in which they will be sold, generally in peat with a nutrient solution.

Purslane (Portulaca oleracea)

Botanical Name: *Portulaca oleracea* L.

Family: Portulacaceae

Common Names: English: purslane, purslave, pursley, pusley; Spanish and Catalan: verdolaga, verdalaga, buglosa, hierba grasa, porcelana, tarfela, peplide (Spain), colchón de niño (El Salvador), flor de las once (Colombia), flor de un día, lega (Argentina); Portuguese and Galician: beldroega, bredo-femea, baldroaga; Basque: ketozki, ketorki, getozca; French: pourpier, portulache

Origin of the Name: The diversity of names and meanings already gives an idea of the age and geographical dispersion of purslane's cultivation or use. On the basis of historical, archaeological and linguistic documentation, De Candolle thought that this species was cultivated more than 4000 years ago. Its common names come from different roots: *lonica* or *louina* (Sanskrit), *koursa* (Hindustani), *kholza* and *perpehen* (Persian), *adrajne agria* (Greek), *portulaca* (Latin, which means "little door", because of the way its capsule opens). The Arabs in the Middle Ages called it *baqla hamqa'*, which means "mad" or "crazy vegetable" because of the fact that its branches spread over the ground without any control. The Hispano-Arabs of Al-Andalus (from the tenth to fifteenth century) used the name *riyla*, which means "foot", most certainly because of its dactyliform leaves, and also *furfir, farfan, farfag, farfagin*, derived from the Persian *perpehen*. They also called it *missita*, which means "mixed", because it is sometimes found growing in gardens and sometimes growing wild. In Spanish, names such as verdilacas, yerba aurato and yerba orate are known (which again mean "crazy herb").

Properties, Uses and Cultivation

As a medicinal plant, it is considered to have antiscorbutic, diuretic and cooling properties. Being rich in mineral salts and with a high water content (95 percent) and mucilage content, it has emollient and soothing properties for irritations of the bladder and urinary tract. It is also used to regulate the bowels. Dioscorides already recognized its medicinal powers: these were anti-inflammatory (eyes) and analgesic (headache), emollient and soothing, antifebrifuge (in juice) and anthelmintic. He also says that "it reduces the desire to fornicate".

In the latter sense, other authors also mention its anaphrodisiac powers (1837 Codex of the Spanish Pharmacopoeia), including this plant among the "four cold seeds", together with chicory, endive and lettuce. The anaphrodisiac effect is perhaps due to the presence of norepinephrin, a precursor of adrenalin, which causes a reduction in the blood flow through constriction of the main arteries. It is also mentioned by Maimonides. In the Middle Ages, the pharmacists of Cairo used to sell purslane seed for various uses, recommending it in particular as a vermifuge. Laguna and Leclerc also recognized its different medicinal properties, especially the anti-inflammatory ones, in mixtures prepared with plantain, violets and gourds. Its magical powers have also been mentioned, as a charm against evil spirits and for dispelling nightmares if placed in the bed.

However, in addition to its medicinal powers, it is also a vegetable, a weed and a food for pigs.

Columela writes in his poem on the garden: "Already the juicy purslane covers the dry beds"; and in *Los doce libros de agricultura*: "Leafy purslane appeases the plot's thirst" (Book X); and in Book XI he gives a recipe for preserving it in vinegar and salt. Paladio refers to it exclusively because of its mucilaginous, medicinal and veterinary properties. Similar references are found in Kastos, taking up the Byzantine tradition. Isidoro de Sevilla mentions it without giving any information on its cultivation. In short, such a summary reference to the Hispano-Roman and Hispano-Visigoth tradition regarding purslane is surprising.

It is the writers of oriental and Arabic treatises who concerned themselves most with this vegetable. Ibn Wahsiyya describes its cultivation in the Near East, presenting it as a summer crop. Most of the Hispano-Arab agronomists deal with this plant. Arib (tenth century) mentions it in his *Calendario agrícola*. Al Zahrawi and Ibn Hayyay (eleventh century) also mention it. Ibn Bassal (eleventh century) deals extensively with its cultivation, already recognizing a certain intraspecific variability (he distinguishes early and late varieties), setting out its temperature and water requirements (summer cultivation and irrigation or vegetable garden), drawing up a sowing calendar which extends from March to August and demonstrating the practice of two basic cultivation periods, depending on whether the aim is to produce seed or to produce for human consumption. Sowing quantities and manuring and irrigation requirements also appear and are dealt with in great detail by the author. Ibn Wafid (Hispano-Arab agronomist of the eleventh and twelfth centuries) mentions it under the names *haqla hamqa'* and *missita*. Ibn al-Awwam, in his *Kitab al-Filaha*, recalls that it is mentioned by almost all the Arab authors and refers to different varieties. He uses the adjectives "mild", "vain" and "blessed".

After the sixteenth century, cultivation of purslane was gradually lost in Spain. Alonso de Herrera (sixteenth century), for example, makes no reference to it while Boutelou and Boutelou (1801) say that "purslane, which is not at all appreciated in Spain, is one of the crops which, in England and other countries further north, need to be cultivated in frames and hotbeds in order to bring forward their vegetation artificially"; and further on: "on this land, it is not usual to cultivate purslane other than using those that have grown at random among other plants cultivated with more care". In spite of Spanish disregard for this plant, it is still valued in many Latin American countries where it was introduced.

Purslane has been eaten as a vegetable, particularly fresh. In England in the seventeenth century, the cooks of Charles II used to add its leaves to all salads, perhaps to satisfy the king's taste or else for its digestive properties. In this recipe, the chopped young leaves were mixed with double the amount of leaves of lettuce, chervil, borage flowers and marigold petals, the mixture being dressed with oil and lemon juice. The recipe resembles that mentioned by Tirso de Molina: "I will have green coriander, garden cress, purslane, borage and mint added to it."

Not only the leaves, but also the stems and rootless plantlets can be eaten raw and fresh. Columela mentions their being eaten pickled with salt and vinegar. Purslane has a pleasant acidic flavour and is very juicy. In Spain, it is usually eaten at a more advanced stage of growth, after cooking. It is also delicious boiled and in omelettes. Sauteed in butter or fried, it is used in soups, broths, salads and sauces. Together with sorrel, it forms part of the French soup *bonne femme*. Recipes are also known for purslane and pea soups. To complete the range of its applications, one could mention its use as an insecticide, in which case its juice is poured on to anthills, and also its ornamental use in Roman and medieval gardens.

At present in Spain, it is basically a volunteer species (weed) among summer irrigated crops, and its consumption is gradually declining; this is also the case with individuals collected from wild populations.

Botanical Description

Purslane is an annual, herbaceous plant, with branched, decumbent or fairly ascending stems of up to 50 cm, and which are reddish, fleshy and glabrous. The leaves measure 0.5 to 3.3 x 0.2 to 1.5 cm, are obovate, entire and fairly papillose. The flowers are yellow and solitary or in axillary groups of two or three. The fruit is in a capsule (pyxidium) of up to 7 mm. The seeds measure 0.6 to 1 mm; they are reniform, black, and maintain their germinating capacity for eight to ten years. Of orthodox behaviour in germination, their viability is maintained much more if they are stored dry at a low temperature.

Ecology and Phytogeography

Purslane was one of the most widespread horticultural plants in the Old World since distant times. It was taken to America where it was naturalized, as in Europe, in gardens, among rubble and at waysides. It originates from the region extending from the western Himalayas to southern Russia and Greece. In eastern Asia it does not seem to be spontaneous. In Greece it is spontaneous and cultivated. Vavilov (1951) categorizes it in the Mediterranean countries of the Near East and central Asia as a weed and vegetable.

Nowadays it is distributed over the hot temperate zones of a great part of the world. Together with other species of the genus it occurs as a weed in the majority of tropical and subtropical countries.

It is cultivated in the United Kingdom, the Netherlands and other European countries. It is a popular winter vegetable in northern India. In Spain, it very frequently occurs as a volunteer, but it is very rare as a crop.

Genetic Diversity

Little work has been done on the management of purslane's extraspecific variability. Apparently, without any aim at improvement, protoplast fusion of the genera *Portulaca* and *Nicotiana* has been attempted, and heterokaryons and the first division have been observed, but it is not clear whether multiple divisions occurred. Nevertheless, there is an enormous intrageneric variability. The genus *Portulaca* is cosmopolitan and many species are grown as a vegetable. Thus, *P. afra* Jacq., *P. pilosa* L. and *P. tuberosa* Roxb. in southern Africa and *P. quadrifida* L. in tropical Africa; *P. retusa* Engelm. in North America and *P. pilosa* L. in South America; *P. napiformis* Muell. in Australia; and *P. lutea* Forst in Polynesia. *P. quadrifida* L. is cultivated in many tropical regions.

Within *P. oleracea* and in its wild populations, Danin and Baker distinguish five subspecies (*oleracea, papillato-stellulata, stellata, granulato-stellulata* and *nitida*), on the basis of the seed size and structure of the testa. Recognition of these subspecies is somewhat questionable, especially if we take into account their sympatric character. Generally speaking, the existence of a single *P. oleracea* complex with several varieties is accepted; it includes: var. *oleracea*, which is widespread as a weed; and var. *sativa* (Haw.) Celak, which is cultivated as a vegetable and has a bigger and erect habit. In a chemotaxonomic study comparing proteins and free amino acids, Prabhakar and Ramayya (1988) found that, within the complex *P. oleracea*, the var. *ophemera* is distinct from the var. *oleracea* and *sativa*. In the var. *sativa*, it is usual to distinguish two types which can be differentiated by their colouring: green purslane and golden purslane. However, it seems that colour depends basically on exposure to the sun and is more an environmental than a genetic characteristic. Some markets, such as the French market, appreciate red in particular.

In the commercial catalogues of seed firms, cultivars of this horticultural plant are not usually offered.

Girenko (1980) has described the intraspecific diversity and composition of cultivars in various climatic zones of the CIS, along with another set of data of agricultural interest.

Extensive work also has to be done on the recovery and conservation of purslane germplasm. In 1985, as part of a joint project with the IBPGR, a mission of the Agricultural Research Corporation collected indigenous germplasm of *P. oleracea* in the northeastern region of the Sudan. At ARARI in Izmir, Turkey, some accessions of *P. oleracea* are conserved.

Cultivation Practices

This is a vegetable which develops rapidly in hot environments. Cultivation is very simple, entailing the necessary hoeing and irrigation on light, rich soils which encourage emergence.

It can be grown in greenhouses and may be broadcast or sown by burying the seeds with light pressure. A first and second irrigation are essential and must be carried out either by sprinkler or by hand. In order to ensure moisture during emergence, the plots are sometimes covered with wet sackcloth. The seeds germinate quickly and have to be raised up to accelerate emergence and development. The plantlets are harvested when four or five leaves have formed which, with suitable temperatures, is achieved in about 20 days. It is possible to cover a long production period by staggered sowing. In temperate areas in central Europe around April, when the frosts are over, cultivation also takes place in the open air with direct broadcasting (10 g per m^2). Moisture must be ensured during emergence. Later, when the seedlings have reached the mid-point in their growth, they tolerate water shortages well. In this type of cultivation, the plant is normally allowed to develop and the stalks are harvested throughout the summer. If the plant is not pulled up, it sprouts again.

The crop's biggest enemies are low temperatures and weeds, which require as many hoeings as necessary. Pests and diseases do not appear to constitute important limitations.

Prospects for Improvement

Cultivation does not present any technical difficulty preventing restoration of this vegetable's use. In experimental tests carried out by the authors on the southeastern coast of Spain, uniform production of seedlings of between 6 and 8 cm was obtainable after a month or so during the winter and spring in an unheated polyethylene greenhouse.

This type of cultivation is the one which may be most readily acceptable on western markets, provided clean rootless seedlings are offered, appropriately packaged in trays covered with plastic film. Under these conditions, they keep well at low temperatures for a couple of weeks.

This type of product is practically unknown to the consumer and yet it is the most suitable for salads. If plants or shoots of plants developed under high temperature conditions are used, they may have excessive mucilage and an unpleasant texture. The plantlets have a milder flavour and texture which make them more appetizing. Where plant material is concerned, practically everything remains to be done, since very little improvement work has been carried out recently.

Borage (Borago officinalis)

Botanical Name: *Borago officinalis* L.

Family: Boraginaceae

Common Names: English: borage, cool tankard; Spanish: borraja, borraja comun, borraga, borracha, bora, corrago, alcohelo, flores cordiales; Catalan: borratja, borraina, pa-i-pexet;

Basque: borrai, borroin, murrum, assunasa, porraina; Portuguese and Galician: borrage, borragem, erva borragem, borraxa

Properties, Uses and Cultivation

Borage is attributed with sudorific (flowers), diuretic (leaves and petioles) and emollient properties (cataplasms of leaves). It contains substantial mucilage, tannin, potassium and magnesium salts and traces of essence. The seeds contain up to 23 percent linoleic acid. Pharmacologists in past times used to include borage within the "four pectoral flowers", and it was also strongly recommended in cases of rheumatism, in which case the fresh leaves were applied as a poultice, since they lose their properties when dry. The flowers and seeds had a reputation as euphoriants and were added to wine for this purpose. Some authors think that borage is the plant which the Greeks called *eafrosinon* and which, according to Pliny, "made men joyous and happy". One Greek proverb used to say: "I, borage, always give courage." In sixteenth-century Spain, it was still attributed with this property.

Thus, Alonso de Herrera (1981 [1513]) states that borages "are healthier than any other vegetable and in truth, it can be said that in many cases they are not appreciated because these powers, which are many, are unknown". He also mentions some of these: "When raw, they engender a very singular blood, and more so when cooked with a good mutton or capons, and for this reason they are very good for old people... and if their seed is drunk in wine, it cheers the heart greatly...". The question arises as to whether the vegetable's virtues might not be due to the other ingredient which accompanied it.

In actual fact, its effects cannot be very obvious since "in many cases they are not appreciated". The mildness of its action perhaps explains the well-known Spanish expression "it is borage water" to indicate that something has come to nothing. For example, Boutelou and Boutelou (1801) explained: "In ancient times it was very often used in medicine, but nowadays it is practically forgotten since it does not produce the effects for which it was applied in those days."

As a food vegetable, the origin of its cultivation has not been pinpointed. Although it is unclear whether the Greeks and Romans made medicinal use of this plant, it is more certain that they did not cultivate it, since none of the writers of treatises such as Columela or Paladio referred to it, although some authors attribute a Latin etymology to *borago* (derived from *borra* = rigid hair, because of the characteristic hairiness of the whole plant). Other authors support an Arabic etymology, from *abu* = father and *rash* = sweat, because of the sudorific property of its flowers. Some historians even thought that the plant came from Africa during the Middle Ages. However, there is no doubt that the plant is native to Spain and that, around the twelfth century, the Andalusian Muslims were not growing it. Indeed, in his *Kitab al-Filaha*, Ibn al-Awwam makes a single reference to it, treating it as a wild plant which could be used in times of famine. Other Andalusian agronomists and doctors such as Ibn Hayyay (tenth century), Ibn Wafid (eleventh to twelfth centuries) and Maimonides (tenth century) seem to mention it, but there is a degree of confusion regarding its name, *lisan al-lawr* (ox tongue), which may refer to both *Borago officinalis* and *Anchusa officinalis* or *A. italica*.

Consequently, borage must not have been cultivated until after the twelfth century. It is known to have been grown in Castile in the fifteenth century and, in 1539, Alonso de Herrera gave an extensive description of its cultivation and properties. It was one of the first vegetables taken to America by the Spanish; as early as 1494 it was being grown in the gardens of La Isabela, the first city founded on American soil. In the seventeenth century, Cobo (1953 [1662]) also stated that borage had adapted to Latin America. In the eighteenth century, it was frequently grown but had already lost importance.

Borage is grown for its leaves and stalks which are eaten as a vegetable. The young leaves can be eaten raw in salad dressed with olive oil, giving an aroma and flavour similar to cucumber. They should be chopped, since they are not very appealing whole because of their hairiness. They are used cooked in soups, as a garnish for meats and also in olla, a kind of stew. The leaves cooked in batter and served with hot or grated cheese are delicious. Similarly, borage dumplings can be made, while its finely chopped leaves can be cooked with almond milk to make an exquisite soup or used to make an excellent borage omelette.

However, nowadays leaf petioles are the part of the plant most used and lend themselves to most of the uses stated.

The flowers are used to garnish dishes and prepare an exquisite dessert. Genders (1988) suggests a recipe for borage tart. In some regions, a dessert is also prepared by frying the leaves, to which sugar or honey is added, in the same way as the *paparajotes* of Murcia, but using borage instead of lemon leaves. In Majorca, according to Font Quer (1990) the leaves are used to make fritters by preparing a mixture with beaten eggs and wheat flour and then frying the leaves thus coated in hot oil and sprinkling them with sugar and cinnamon.

Borage is also a honey-producing plant, the flowers and roots produce dye, while the active synthesis of linoleic acid—of pharmacological and cosmetic interest—occurs in the ovary, which explains the high content of linoleic acid in the seeds.

Botanical Description

Borage is a sturdy, annual herbaceous plant. Almost all the plant is covered with stiff hairs. It has a taproot and erect, sturdy stems which reach 20 to 100 cm and are sometimes branched. It has ovate or lanceolate, petiolate basal leaves in a rosette which grow up to 25 cm. The upper caulinar leaves surrounding the stem are sessile. The flowers are a bright celestial blue on branched tops. Flowering occurs from spring to autumn. The fruit contains four oblongo-ovoid nucules measuring 4 x 2.5 mm.

Borage is an allogamous plant, which has hermaphrodite flowers with exserted stamens. It has a self-incompatibility system controlled by numerous genes. Pollination is predominantly entomophilous (bees).

The plant is propagated from seed. Seed collection is laborious, since the seeds drop easily. Sixty-five seeds weigh 1 g; 1 litre of seeds weighs around 430 g. In commercial storage conditions, germination capacity remains high for eight to ten years. Its behaviour is orthodox in storage. The seed germinates very quickly, without any dormancy problems. The chromosome pattern is 2n = 2x = 16.

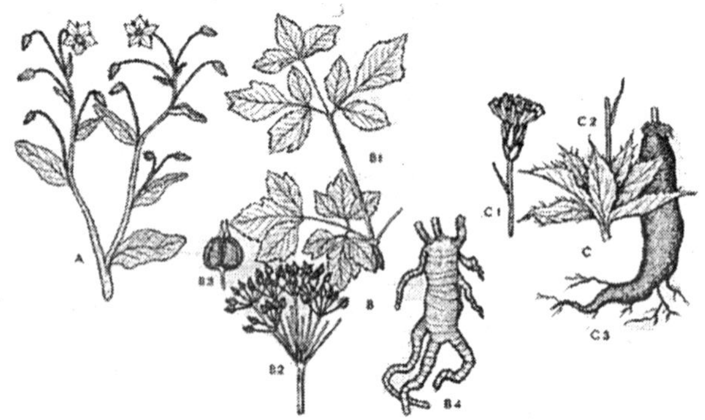

Horticultural crops: A) borage (Borago officinalis); B) alexanders (Smyrnium olusatrum); B1) leavès; B2) inflorescences in the umbel; B) fruit; B4) root; C) scorzonera (Scorzonera hispanica); C1) capitulum; C2) basal rosette of leaves: C3) root

Ecology and Phytogeography

In its spontaneous or subspontaneous form, borage grows on uncultivated land, embankments, fallow land, wasteland, garden edges, waysides and among ruins. It is native to the Mediterranean region but has been naturalized in the hot zones of western, central and eastern Europe, sometimes with unstable escapes northwards. It is also found in Southwest Asia, Macronesia and North America. Cultivation of borage as a vegetable is limited to certain regions of the Netherlands, France, Spain and Latin America, being unknown in the rest of the world. In Spain, it is grown mainly in the Ebro valley, in the provinces of Zaragoza, Logrono and Navarra. The total cultivated area in 1987 was 303 ha and production 7818 tonnes.

In recent years, some expansion of cultivation towards Andalusia has been noted, particularly in Almeria. Sheltered cultivation is beginning to be carried out, with excellent results.

Genetic Diversity

The genus *Borago* has only two Mediterranean species. In humid areas of Corsica and Sardinia, *B. pygmaea* (DC.) Chater & W. Greuter, a perennial with decumbent stems, is found.

Borago officinalis L. is a very variable species. There are varieties characterized by the flower colour. Although they are generally bright blue, there are also types with white and pink flowers. However, these are very heterogeneous populations with a great diversity in habit, vigour and development of the plant, shape, colour and size of the limb and leaf petiole, flowering, etc. The cultivar Flor Blanca, which is marketed in Spain, has leaves with petioles of 40 to 50 cm in length and 1.5 cm in width. The plant grows to a height of around 50 to 60 cm. In the gene bank of the SIA at the Diputacion General de Aragon (Zaragoza), there is a small collection of accessions of this vegetable.

Cultivation Practices

Borage is a very hardy plant which is suited to all types of soil, although it grows best on clayey-muddy soils. It prefers land that is rich in organic matter. It tolerates low temperatures, down to -50°C, and starts to sprout again when the temperature rises.

In Spain, direct sowing is used. The ground should be prepared with a basal dressing using about 50 tonnes of manure per hectare, if it has not been incorporated into the previous crop, and 90 to 120 units per hectare of nitrogen, phosphorus and potassium. The soil must be well broken up with deep ploughing and a couple of harrowings. In Aragon, staggered sowings are carried out in the open air from mid-August to January, in rows or individual drill holes with 25 to 30 cm between plants.

Cultivation presents no particular problems; the plants must be irrigated and, in the event of intensive cultivation, after thinning out top-dressing must be supplemented by 150 units per hectare of easily assimilated nitrogen.

The vegetative cycle takes between 50 and 120 days and harvesting can begin in mid-October, ending in May since, when high temperatures come with spring, the plant goes into flower and loses its value. Harvesting is done by hand. Each plant has two or three rosettes with five to seven leaves each, with a weight of 500 to 1000 g per plant.

Production levels of around 60 to 100 tonnes per hectare are obtained. According to data in the Spanish Government's *Anuario de Estadistica Agraria*, average yields are 25 tonnes per hectare in the case of open-air irrigation and 36 tonnes per hectare in sheltered cultivation, Navarra being foremost with yields of 40 tonnes per hectare using both methods of cultivation.

Recently, sheltered cultivation under plastic has been gaining in importance. Under these conditions, much longer and fleshier leaf stalks are obtained and the stalk/plant yield rises to 60 percent, as against the 40 percent obtained with open-air cultivation. Production levels are also usually better.

The crop's main enemies are virus diseases (cucumber mosaic virus), soil fungi (*Fusarium* sp.), soil grubs, caterpillars and aphids.

The plant is usually marketed in 15 to 20 kg "bundles", amounting to 15 to 30 clumps, or in 10 to 12 kg boxes as complete plants, with part of the leaf removed. However, the consumer prefers borage to be completely stripped and packed in trays protected with plastic film.

Borage is subject to the technical regulations on the control and certification of horticultural plant seeds. The requirements for seeds of the basic, certified and standard category are 97 percent specific purity, 65 percent germination of pure seeds, with a maximum tolerance of 0.5 percent of seeds of other species. According to INSPV data, in 1989 2567 kg of borage seed were marketed, 2 489 kg of which were homegrown. Only the white variety was grown.

Another method of cultivation carried out in the Netherlands uses plantlets. After direct sowing, these are allowed to grow to a height of 10 to 15 cm and the complete plantlets are harvested. After washing and root removal, these can be marketed in trays covered with plastic film.

Prospects for Improvement

Most improvement work has been carried out using white flower types. Breeding by growers has created forms with more succulent, longer and wider leaf stalks, with little pigmentation and less hair than the wild forms.

One of the main problems of cultivation is its ease of bolting, including the formation of flowers, which lowers the value of production. This process is caused by high temperatures and light intensity and reduced humidity. Breeding for resistance to bolting is a priority improvement objective, and a very high response to breeding is observed. Although this plant has traditionally been cultivated in the open air, excellent results are now being obtained under plastic, in which case growth improves. A quality product, with long, tender leaf stalks and less hair can be obtained for a good part of the year in a greenhouse. The plant tolerates low winter temperatures and high humidity well. In the area around Zaragoza, borage has been converted into the most profitable crop under plastic.

The expansion of sheltered cultivation may encourage the recovery of this marginalized vegetable. The first tests in this connection have been carried out in Almeria. If they prove positive, they would contribute to the diversification of production and to improving the supply in this region, which has great agricultural importance and yet depends on a very small number of crops.

As far as the consumer is concerned, in the case of regions that do not have a tradition of using this plant, borage must be presented stripped and properly packed, so that the work of culinary preparation is reduced. The plant's coarse, hairy appearance may cause some degree of rejection, which is avoided with appropriate cleaning and presentation.

With sights set on possible external markets which are even more demanding than the Spanish market, the high nitrate content of leaves and leaf stalks will need to be reduced. This can be achieved without great difficulty, as breeding to obtain a low nitrate content has been effective in other cases. Breeding to obtain individuals with a low content of lasiocarpine, a pyrrolizidinic alkaloid, would also be advisable, although its content is not excessively high.

As regards the plant's pharmacological use, in vitro cultivation of embryos is being developed; this is a technique whereby the active synthesis of linoleic acid takes place. In vitro propagation techniques of borage have also been developed.

Alexanders (Smyrnium olusatrum)

Botanical Name: *Smyrnium olusatrum* L.

Family: Apiaceae = Umbelliferae

Common Names: English: alexanders, alisander, maceron; Spanish: apio caballar, apio equino, apio macedonico, perejil macedonico, esmirnio, olosatro, canarejo; Portuguese and Galician: salsa de cavalo, cegudes, apio dos cavalos, roses de pe de piolho; Catalan: api cavallar, abil de siquia, julivert de moro, cugul, aleixandri

Origin of the Name: This is the *hipposelinon* of the Greeks, a word which means parsley or "horse celery". In Arabic, during the Andalusian period, it was called *karafs barri*, one of the various *karafs* (celeries) known by Hispano-Arab agronomists, different from cultivated celery (*Apium graveolens*), aquatic celery (*A. nudiflorum*) and mountain or rock celery (the Greek and Latin *petroselinum* or *oreoselinon*). Alexanders has always been identified as oriental or Macedonian, very possibly as a reference to its geographical origin and its allochthonous character.

Properties, Uses and Cultivation

Its use as a medicinal plant is very old. The Greek botanist Theophrastus (fourth century BC) made reference to the plant. Dioscorides (first century) also included it in his *Materia medica*, commenting that its roots and leaves were edible. According to this author, its seed, taken with wine, is an emmenagogue. However, Galen said that it was less active than celery. In the Cordoba of the caliphs, Maimonides also spoke of its powers. During the Middle Ages, it was constantly considered as a plant with diuretic, depurative and aperient properties, particularly through its root. However, its most outstanding quality was perhaps as an antiscorbutic because of its high vitamin C content. The fruit has carminative and stomachic properties. In the eighteenth century, it continued to maintain its reputation as a medicinal plant, as the *Flore economique des plantes qui croissent aux environs de Paris* described it in 1799.

The plant, and especially the leaves, have a smell and flavour similar to myrrh. Hence the origin of the word *smyrnion*, its generic name. Columela (first century) refers to the plant as "myrrh of Achaea", because it was grown in Greece, which the Romans called Achaica or Achaea. It is also because of its characteristic flavour and smell that it is used as a condiment; it is used to season food in a similar way to parsley, giving flavour to soups and stews, and to prepare sauces accompanying meat and fish. However, its commonest use has been as a fresh vegetable, with a preference being shown for its leaves, young shoots and leaf stalks, which impart a pleasant flavour similar to celery, although somewhat sharper. It has also been eaten cooked. The Latin word olusatrum, which means "black vegetable" reflects these uses. The roots were used preserved in a sweet-and-sour pickle. The fruit contains an essential oil, cuminal, which is reminiscent of cumin.

The history of its cultivation is surprising. Of all the Umbelliferae used as vegetables, alexanders has been one of the commonest in gardens for many centuries, although in the nineteenth century it was almost completely forgotten. It was probably being gathered before the Neolithic period and was already being grown as early as the Iron Age. It became very popular during the time of Alexander the Great (fourth century BC) and was widely grown by the Romans, who certainly introduced it into western and central Europe, including the British Isles. It is now naturalized in these regions and on the Iberian Peninsula.

Columela elaborates on its cultivation and methods of consumption: "Before alexanders puts out stems, pull up its root in January or February and, after shaking it gently to remove any soil, place it in vinegar and salt; after 30 days, take it out and peel off its skin; otherwise, place its chopped pith into a new glass container or jar and add juice to it as described below. Take some mint, raisins and a small dry onion and grind them together with toasted wheat and a little honey; when all this is well ground, mix with it two parts of syrup and one of vinegar and put it like this into the aforementioned jar and, after covering it with a lid, place a skin over it; later, when you wish to use it, remove the pieces of root with their own juice and add oil to them."

Isidorode Sevilla (sixth century [1982]) seems to attach less importance to alexanders.

In France, it was an important vegetable, and was grown on the estates of the Carolingian kings. Thus, in the *Capitular de Villis*, promulgated by Louis the Pious, son of Charlemagne (around AD 795), alexanders appears among the plants which should be cultivated. In the eighteenth century, in Versailles, it was used blanched to accompany winter salads. In the early nineteenth century, Rozier, in his *Dictionnaire universel d'agriculture pratique*, writes: "The leaves of alexanders can appear among cooking condiments, like parsley. Its roots and young shoots are still eaten in England after blanching in the same way as celery."

There is documentation on its cultivation in Belgium in the fifteenth century and on its abundance in English gardens in the sixteenth century. The Italians also traditionally used this plant. However, by about the eighteenth century its cultivation was only very occasional or had fallen into disuse. In Spain, Font Quer (eighteenth century [1990]) says that its root was eaten in many countries as a salad, raw and cooked, as were the stems and young leaves. By the nineteenth century, Spanish agronomists were no longer making any reference to it. Thus, Boutelou and Boutelou (1801) do not mention it, an omission which contrasts with the 13 pages devoted to celery cultivation.

Alexanders was falling into disuse as from the seventeenth century, in direct competition with the celery of the Italians", an improved form of wild celery (*Apium graveolens*). This is a case of marginalization in which one plant, doubtless widely used since prehistory, is replaced by another one improved later.

Botanical Description

Alexanders is a biennial herbaceous plant with a thick elongated root. The stems grow up to 150 cm and hollow on fruiting. It has large, pinnatisect, basal leaves, with ovate to subrhombic terminal segments; the caulinar leaves are pinnatisect. The umbels have seven to 22 rays, with black, didymous fruit measuring 5.5 to 7.5 x 4 to 7.5 mm. Alexanders flowers from April to June and propagates well from seed. Its chromosome structure is 2n = 2x = 22.

Ecology and Phytogeography

Wild populations of alexanders grow abundantly in salt-marshes and uncultivated land near the sea, normally in lime soils. It is also found in hedges, woods and on waysides.

It is spontaneous throughout southern Europe, North Africa (Algeria) and in the Near East. In former times it was very abundant in the area around Alexandria. Vavilov (1951) places this crop in the Mediterranean gene centre.

It also occurs on the Canary Islands and in the rest of the Macronesian region.

Genetic Diversity

Perfoliate alexanders (*Smyrnium perfoliatum* L.) has smaller fruit (3.5 mm long) and is distributed through central and southern Europe and southwest Asia. The blanched stems and leaves are used in salads. Its cultivation is documented in the sixteenth century. According to Mathon (1986), this species is of superior quality.

Nowadays it is very difficult to find cultivars of alexanders. However, several cultivated varieties must have existed. For example, in England in 1570, Petrus Pena and Mathius Lobel wrote: "...the cultivated form is far better than the wild plant...". It seems that the plant is still occasionally grown in Great Britain.

Accessions of this species are kept only in the gene bank of the Córdoba Botanical Garden. They are from wild populations in Andalusia.

Cultivation Practices

According to Columela, "alexanders must be grown from seed in ground dug out with a *pastino*, particularly close to walls because it likes shade and thrives on any kind of ground: so once you have sown it, if you do not uproot it fully but leave its stems for seed instead, it lasts forever and requires only light hoeing. It is sown from the feast day of Vulcan (August) until the calends of September, but also in January...".

Nowadays, since cultivation has been relegated to a few family gardens, similar practices are frequently seen. The stem is left to seed, and sowing and spontaneous cultivation takes place. Something like this usually occurs with chard: weeds are removed and a little fertilizer is applied.

Modernization of this crop will depend on techniques similar to those used for celery, including blanching, taking into account the fact that alexanders requires less soil and water.

Prospects for Improvement

Celery was also known from antiquity but was considered to be an inedible plant of ill omen. The Greeks, who called it *apion*, used it in funeral ceremonies. It appears to have been grown early in our era by the Latin's. Columela refers to it:...after the ides of May, nothing must be put in the earth when summer approaches, except for celery seed, which must nevertheless be watered, since in this way it does very well...". Paladio also mentions it, probably basing himself on the earlier source. Likewise, in the *Capitular de Villis* (eighth century) reference is made to both *apium* and *olisatum*. Throughout this period, cultivation of alexanders seems to be predominant.

Around the seventeenth century, types of celery appeared which were derived through breeding to obtain a better size and improved succulence of the leaf stalks (var. *dulce* (Mill.) Gaud.-Beaup.) or fuller leaf development (var. *secalinum* Mill.) and which were clearly differentiated from the wild plant. These types are actually different vegetables requiring specific cultivation practices. Thus sweet-leaved celery ("celery of the Italians") is well suited to "blanching", which enables a milder, more tender product to be obtained.

The marginalization or disuse of many vegetables used since ancient times in Europe may be connected with the changing tastes in the Western world. The trend has been away from dishes rich in spices and hot ingredients towards milder dishes, which respect the flavour of the food itself or enhance it. This is perhaps the case with celery *vis-a-vis* alexanders. Alexanders is more bitter and pungent and not as tender as sweet celery.

It is significant in this respect that the last agronomic references to the cultivation of alexanders mention the introduction of the blanching technique. It appears thus in the reports by Versailles and Abbot Rozier: "...after they have been blanched in the same way as celery..."; and Barral and Sagnier, in *Diccionario de agricultura* (1889), write: "...in Turkey the cultivation of this plant is still an honour. The leaf is eaten after it has been blanched...". The blanching technique also used to be employed in North America. It is obvious that the smaller plant, celery, had asserted itself and now served as a reference, making it necessary to adopt the same cultivation practice for alexanders, evidently with little success.

While cultivation of alexanders is waning, cultivation of celery is by contrast on the increase, as is its importance in cool subtropical and tropical areas of Latin America and the Far East. Petiolate cultivate with big leaves are chiefly used.

The recovery of alexanders would be achieved via the derivation of plant materials with a specific typology, for specific uses, and the development of associated agronomic techniques; this seems very unlikely.

Scorzonera (Scorzonera hispanica)

Botanical Name: *Scorzonera hispanica* L.

Family: Asteraceae

Common Names: English: scorzonera, Spanish salsify, black oyster plant, viper's grass; Spanish: escorzonera, escorcionera, escurzo, yerba viperina, salsifi negro, salsifi hispanico, churrimana, tetas de vaca; Catalan: escurconera; Basque: sendaposei, astobe-harri; Portuguese and Galician: escorcioneira, escorzoneira.

Properties, Uses and Cultivation

Scorzonera has diuretic and depurative properties. The root has restorative and sudorific properties and is an ingredient of many infusions. It is very rich in carbohydrates (18 to 20 percent in fresh weight), with a high proportion of inulin and laevulin, which makes it very suitable for a diabetic diet. It also contains conopherin (glucoside), asparagine, arginine, histidine and choline. In upper Aragon, the latex is added to milk as a cure for colds. Its ground, fresh leaves are used against viper bites to soothe the pain. Its peeled root, fresh or cooked, acts as a tonic for the stomach and fortifies the body.

It is considered to be an antidote to the bite of poisonous animals, for which reason in Spanish it is called "escorzonera", *i.e.* herb against *escuerzo* [toad]. The *Diccionario de la 'engua espanola* of the Real Academia Espanola mentions that the name derives from the Latin "black root" because of its external colour. In Italian, too, *scorza* means "root" and *nera* "black". However, as documented in Mattioli's *Epistolarium medicinalium libri quinque*, published in 1561, the first interpretation seems correct.

Cultivation of this plant is thought to be recent. No Roman or Arab agronomist mentions it. In Spain, its cultivation is not dealt with either by Andalusian agronomists (tenth to fourteenth centuries) or Castilian writers of treatises in the sixteenth century. The same applies in other countries. In France, it is not mentioned in the *Capitular de Villis* of the Carolingian kings, nor does Olivier de Serves, Henry IV's minister, mention it. It was from the sixteenth century onwards that botanists began to concern themselves with this species, describing it as wild, although sometimes introduced into botanical gardens. It is not quoted as a cultivated plant until up to one century later. In time, it was to become fashionable in several countries. Thus Louis XIV of France was very fond of it.

Although scorzonera was perhaps first cultivated in Spain, its cultivation has never been very important in the country. Boutelou and Boutelou (1801) commented: "Scorzonera is usually sown on the edges of unoccupied beds, the empty spaces being profitably used by this tasty root", thereby demonstrating a marginal rather than a main crop.

On the other hand, it is curious that these same authors visualized a greater agricultural importance for white salsify, than scorzonera, contrary to what actually happened. Thus, they thought that "...sometimes the roots of scorzonera can begin to be used the first year after sowing, but they are so thin that there is no point in wasting them so young. They require two or sometimes three years for their root to form. Salsify, which has the same taste and properties and which forms in one year, should be preferred because it requires less time in the ground and its product is much more plentiful." The main improvement activity on this crop has enabled some good cultivars to be obtained, with a greater growth rate and better yields than salsify in annual cultivation.

The part of the plant most used is the tender, fleshy root. It is peeled and then cut into pieces and placed in water with lemon to prevent it from turning black. It can then be eaten in a wide variety of exquisite dishes: raw in a salad; dressed with vinaigrette or with other sauces, steamed and served with Bearnaise or Bechamel sauce or with whole milk cream and toast; sauteed in butter with parsley or other herbs; boiled as an accompaniment for meat; grated with cheese; baked with tomato and roast mutton or pork, fried with oil or butter after being lightly cooked and served with lemon; scrambled with eggs or in omelettes; and preserved in sugar.

It is recommended that, once cooked, the roots should be peeled so that they do not lose their flavour.

The leaves can also be eaten, especially the young ones after boiling. The "beards"—young, fresh and tender leaves—can also be eaten raw. The young shoots are used in the same way as asparagus. The flowers are added to salads as a flavouring. They have an aroma reminiscent of cocoa. For this purpose, the flowers of other species such as *S. mollis* and *S. undulata* are also used. The flower buds can be used too. Recipes exist for scorzonera flower omelette.

Botanical Description

Scorzonera is a perennial plant with a long, fragile taproot, which is blackish on the outside and white and milky inside, and which increases in size each year. The stems are solitary or few in number, usually branched on the upper part and between 30 and 120 cm long. The leaves are broad, long, fleshy and spatulate. The yellowish flowers are in capitula at the end of the stems. Flowering is in spring and summer (April-June).

Propagation is from seed. The achenes are 10 to 20 mm long, cylindrical, whitish and rough, with a pappus that has several rows of hairs. The weight of 75 to 90 seeds is I g, the weight of one litre of them is around 580 g. Under ordinary storage conditions they maintain a high germination capacity for two to three years.

It has a diploid chromosome number: 2n = 14. In the var. *crispatula*, some polyploids have been detected: 2n = 4x = 28.

Ecology and Phytogeography

Scorzonera grows on dry pasture, rocky areas, in thickets and on limy or marry soils of temperate zones.

It is distributed over central and southern Europe and the south of the CIS, although it is not found in Sicily or Greece or in northwestern Africa or southwest Asia. It probably originates from the Mediterranean region and is native to Spain.

The plant is little cultivated outside Europe. Most cultivation takes place in the gardens of amateurs, with the plant being cultivated in professional gardens on a very small scale. Some estimates put cultivation at only a few dozen hectares. The countries with the biggest cultivated area of scorzonera are Belgium, Poland and members of the CIS.

At present, its cultivation is practically unknown in Spain. Although it is subject to the Technical Regulations on the Control and Certification of Agricultural Seeds and Plants, there is no evidence of the seed being marketed in Spain in recent years.

Genetic Diversity

The modern *Scorzonera* genus, which is very close to *Tragopogon*, only includes three sections (*Podospermum*, *Scorzonera* and *Lasiospora*) with some 28 species in Europe. The majority of them are perennial diploid plants with 2n = 2x = 14. Cytotypes also exist with 2n = 2x = 12, x = 6 being derived from the earlier type through translocation.

In Spain, some 13 species are to he found. The majority of them prefer dry soils. This is the case with *S. angustifolia* L.,

S. transtagana Coutinho, *S. hirsuta* L., *S. crispatula* (Boiss.) Boiss. and *S. brevicaulis* Vahl. *S. parviflora* Jacq. is found predominantly on saline soils; *S. laciniata* L. on alkaline soils; *S. aristata* Ramond ex DC. is calcicolous and is found only in meadows and other grassy places of the Pyrenees, the Alps and Apennines; *S. fistulosa* Brot. del W. in Portugal and southwestern Spain. *S. humilis* L., dwarf scorzonera, grows very widely in Europe, while *S. baetica* (Boiss.) Boiss., *S. albicans* Cosson and *S. reverchonii* Deveaux ex Hervier are found only in southern Spain.

Scorzonera (*S. hispanica* L.) is extremely variable, especially in its leaf shape. The botanical varieties recognized are *crispatula* Boiss. (*S. crispatula* (Boiss.) Boiss.), which is very widespread, and *pinnatifida* (Rouy) Díaz de la Guardia & Blanca, which is relatively rare; they are basically distinguishable through their leaf morphology.

Numerous commercial cultivars already exist, and there are generally populations with open pollination:

- Gigante de Rusia, with a regular cylindrical, very long and smooth root and a very black skin. Various selections derive from it, such as Gigante negra de Rusia, Gigante anual, Annual Giant Bomba, Russisk Kaempe, etc.
- Lange Jan, which is of good quality.
- Elite Stamm, which is productive, stable, with a high yield of superior size roots.
- Schwarze Pfahl, which is similar to Elite Stamm.
- Pronora, which has well-formed roots, a smooth skin and, when canned, a good colour and flavour. It is especially suitable for industrial processing.
- Vulcan, Duplex and Pilotis, which are suitable for the frozen foods industry.
- Hoffman 83, Flandria, Nero, Duro and Habil are also good cultivars.

There are collections of local races and old cultivars at the Rijksstation voor Plantenveredling de Merelbeke (Belgium), at the Nordic Gene Bank in Alnarp (Sweden) and at the Vavilov Institute of Industrial Plants, St. Petersburg.

Cultivation Practices

Scorzonera is a vegetable that resists drought well when the plant has already developed. It has similar cultivation requirements to white salsify. It is a typically winter vegetable which, although perennial, is grown as an annual.

It is usually sown direct in early spring, in shallow furrows, with 25 to 35 cm x 12 to 15 cm spacing. Care must be taken to provide protection from birds, which are very fond of these seeds.

About 12 kg of seed per hectare is required. Deep, fresh, loose soil is needed; it must be rich in decomposed organic matter and free from stones or gravel, which cause root deformation. The basal dressing recommended is 30 tonnes per hectare of rotted manure, 50 units of N, 100 units of P_2O_5 and 200 to 250 units of K_2O).

Attention must be paid to the first irrigations and hoeings, which can be controlled chemically, both at pre-emergence and post-emergence, with CIPC. It prefers sunny soils and the presence of easily assimilable nitrogen of which an additional 50 units can be applied as a top-dressing.

Harvesting takes place from November to March and requires perhaps more care than the harvesting of white salsify, since the roots are very fragile. This means furrows of about 40 cm have to be opened parallel to the rows of roots. Storage is good, both on the actual cultivation land and in cold stores at between 0 and -1°C, possibly for two to three months, or frozen, with light industrial processing to clean, peel, cut and scald the vegetables to prevent oxidation.

Yields of around 20 to 30 tonnes per hectare have been obtained.

The most important diseases are mycosis, white rust, oidiopsis and strangulation and splitting of the roots, the aetiology of which is unknown.

Prospects for Improvement

Although it is thought that this vegetable is very little cultivated in Spain, because it has not been introduced into

Iberian cooking, it should be recognized that serious cultivation problems still exist.

Although scorzonera is more productive than salsify and its cultivation more frequent, the two crops have many problems in common:

- a prolonged cultivation cycle, with garden space being occupied for an excessively long time;
- susceptibility to bolting, even during the first year of cultivation—although this does not hollow the root or impair its quality, it does affect yield, making systematic cuts of the flower stems necessary;
- poor seed storage;
- slow emergence and the need for a constant level of moisture;
- very laborious harvesting, since deep trenches have to be opened because the roots are very long and fragile;
- high nitrate content.

Some of these problems have already been tackled or are on the way to being solved. Thus, Schwarze Pfahl is more resistant to bolting than Elite Stamm.

Einjahrige Riesen is particularly resistant to bolting and produces a low percentage of roots with cavities. However, it does not attain the yields of the former. Since genetic variability in respect of the character exists within commercial cultivars, rapid progress in improving this cultivar may be expected.

In Belgium, material is being selected which is especially suited to mechanical sowing and harvesting. Lange Jan, Hoffman 83 and Flandria were the ones which contributed the best product qualities among the cultivars tested.

In Poland, work is being done on the development of cultivars suited to industrial processing (both canning and freezing); some cultivars display a good behaviour in this respect.

Insofar as these improvement objectives are achieved, scorzonera may be expected to begin acquiring greater economic importance. It should not be forgotten that it is a vegetable with a very delicate flavour; its glucide composition is rich in inulin, very unlike other tubers and roots rich in carbohydrates, for instance the potato which has a high starch content. This property may be the reason for the increase in demand and price.

Spotted Golden Thistle (Scolymus maculatus)

Botanical Name: *Scolymus maculatus* L.

Family: Asteraceae = Compositae

Common Names: English: spotted golden thistle; Spanish: tagarnina, diente de porro; Portuguese: escólimo-malhado

Origin of the Name and Properties: The generic name derives from the Greek, *skolos*, meaning spines, a characteristic shared with many other Compositae. In ancient Greece, a thistle with an edible root was known by the name skolymos. Diuretic and antisudorific properties were attributed to these plants.

Spotted golden thistle has occasionally been cultivated, but generally the wild plant has been used, with harvesting being limited to the leaves only in spring. At present, its cultivation is very restricted and is tending to disappear.

Cervantes did not seem to set great store by this plant: "...1 do not have a stomach made for spotted golden thistle, nor for *piruetanos*, nor for roots of the forests." However, the fleshy parts of the young leaves, like those of Spanish oyster plant, constitute a delicious vegetable which can be used in soups, stews and scrambled eggs or as an accompaniment for meat. Baked au gratin, they make an excellent dish.

Genetic Diversity

The genus *Scolymus* L. includes another two Mediterranean species with a use similar to that of the spotted golden thistle, the Spanish salsify or Spanish oyster plant (*S. hispanicus* L.), with a wide Mediterranean distribution, and *S. grandiflorus* Desf, with a more restricted distribution in the eastern Mediterranean. These are very close species which differ in the leaf margin and wings of the stem and in the involucral bracts. among other characters. Unlike the spotted golden thistle, these Spanish salsify oyster plants are biennial or perennial.

A great morphological variability is observed, but no collections of material are known.

Cultivation Practices

The spotted golden thistle is a very hardy plant which prefers clayey soils, although it grows spontaneously in a wide variety of environments. It tolerates cold and drought.

The method of cultivation is similar to that of Spanish salsify, although the latter thrives better on looser soils. Sowing is direct into the soil ready for cultivation, in late winter, with furrows 30 cm apart. After thinning, the plants are spaced 30 cm apart. It is preferable to apply organic fertilizer beforehand. The usual cultivation practices are very simple, being limited to removing weeds.

With hot temperatures, the plant grows very rapidly, with the basal rosette forming quickly, at which time the leaves have to be harvested.

Prospects for Improvement

Spotted golden thistles, like Spanish salsify or oyster plants, are practically unknown vegetables on the market. However, they are appreciated in many Spanish regions on account of their very pleasant flavour. As in the case of so many other crops, its revival will have to be accompanied by a marketing system which creates demand. This means publicity campaigns, utilization standards, recipes for traditional dishes, etc., as well as a product of sufficient quality being available on the markets. The fleshy leaf parts would have to be offered peeled and clean and suitably packaged.

From the point of view of improvement, one of the most serious problems of the spotted golden thistle is the ease with which it goes into flower, encouraged by long-day spring conditions and high temperatures. Selection for resistance to this process would increase the cultivation period and make it possible to improve yields of the basal rosette. The plant's general spininess is another problem.

Undoubtedly, the most urgent task is to carry out collecting expeditions in the Mediterranean basin, including the Maghreb, and to characterize the material collected as a starting point for improvement. At the present time it is already very difficult to find traditional cultigens. This problem is not limited to the spotted golden thistle and Spanish salsify, or even to the genus *Scolymus*, but affects many other Compositae. For example, the tribe *Carduaceae* contains 80 genera with over 2650 species, 227 of which are found in Spain and 150 of which are endemic in the country. Many of these plants have agricultural value and have occasionally been cultivated. In the majority of cases, cultivation is on the decline, even though it is being maintained. The recovery of these genetic resources, the characterization of the materials and the initiation of improvement programmes could contribute towards diversification, both of production and supply, thus helping to make Spanish agriculture more competitive.

Spanish Salsify (*Scolymus hispanicus*)

Botanical Name: *Scolymus hispanicus* L.

Family: Asteraceae = Compositae

Common Names: English: Spanish salsify, Spanish oyster plant, common golden thistle; Spanish: cardillo, cardillo de comer, cardillo de olla, cardillo bravio, cardo lechar, cardon lechar, cardon lechal, lechocino, cardo zafranero; Catalan: cardet, cardelina; Basque: kardaberaiakca; Portuguese: cardo de ouro, cangarinha

Properties and Uses

The Spanish salsify plant has been recognized as having antisudorific and diuretic properties. The Greeks knew it and it is mentioned by Theophrastus. Pliny makes reference to it and considers it an antiperspirant. However, it is barely mentioned by Andalusian agronomists. The translator of an anonymous Hispano-Arab document of the eleventh and twelfth century interprets that *silyan* and *adaliq*, spiny plants which people collect among wild vegetables, are indeed Spanish salsify, *Scolymus hispanicus*. Although it has been cultivated occasionally, at present it is clearly in recession. Most of the Spanish salsify that is eaten comes simply from collecting the wild plant.

Several parts of the plant have a fairly delicate flavour. The young basal leaves are eaten as a vegetable in salads, boiled, in soups, stews, omelettes, etc. The most pleasant part of the leaf is the central rib, a white fleshy part which is obtained by peeling the leaf, with a scraping movement with one hand from the base to the apex, while the other hand holds the base. The young stems are used in a similar way. Font Quer (1990) mentioned that this plant is appreciated in almost all of Spain's provinces and "...is used widely in stew during the spring". In the sixteenth century in Salamanca, the washed young plants used to be eaten with their root, either raw or in stews with meat. In soup, its roots are prepared with milk, butter and flour.

Botanical Description

Spanish salsify is a biennial or perennial plant, which is erect, contains latex and is very spiny. The stems are between 5 and 250 cm long, branched at the top, with discontinuous spiny, dentate wings. The basal leaves are oblong-lanceolate, smooth, pinnatisect, with few spines, and a long petiole. The caulinar leaves are rigid, coriaceous and spiny. The capitula have one to three golden-yellow, enveloping leaves; they are about 3 cm long, in a lateral or terminal arrangement and surrounded by an involucre of spiny bracts. The achenes are 2 to 3 mm with a pappus that has a short corona. It flowers from May to July. The plant is propagated from seed, which has a very good germination capacity for several years and does not exhibit any marked dormancy phenomena. It is a diploid plant: $2n = 2x = 20$.

Ecology and Phytogeography

Spanish salsify is found on waste ground and uncultivated land, among rubble, in ditches and along paths; it is most frequently found in sandy places in temperate zones.

Distributed through southern Europe and North Africa, it extends to northwestern France. Vavilov (1951) pinpoints its origin as the Mediterranean region. In Spain, it grows wild in most of the country but shuns high mountains; it is less common in the north. It is also found in the Canary Islands.

It is occasionally cultivated in Mediterranean countries such as Spain, Greece and the Maghreb; it is practically unknown in the United States.

Genetic Diversity

There is considerable variability in the morphological characteristics of Spanish salsify such as hainness, leaf morphology and involucral bracts, re-ceptacular scales, spininess, etc.

No definite cultivars exist; it is still possible to obtain a few cultigens, although there is a serious risk of losing these materials.

There has been no significant activity in collecting or conserving genetic resources of this species.

Cultivation Practices

Spanish salsify is a very hardy plant, is resistant to cold and thrives on all kinds of soil, although it prefers light-textured soils that are rich in organic matter. Its cultivation requires very little care.

Sowing is direct and is carried out in late winter or in spring. A light, well-drained, manured soil should be used. It can be sown in furrows, 30 cm apart with a distance of 30 cm between plants after thinning.

The young white shoots can be pulled up when they reach 20 cm or so in height. The fleshy parts of the leaves need the basal rosette to be well formed. The roots are usually harvested around the end of October or during the winter. If the plant is left until the following year, it goes into flower and develops a sturdy stem, while the basal leaves lose their quality because of toughening. Therefore, although the plant can be kept for several years, it should be cultivated as an annual.

There are no serious phytopathological problems.

Prospects for Improvement

The considerably spiny nature of the Spanish salsify plant, and especially of the caulinar leaves which have big, tough spines, is a serious drawback to its handling and deters attempts to cultivate it. The breeding of less spiny forms would facilitate the plant's handling. As far as the most widely used portion is concerned—*i.e.* the fleshy part of the leaves—forms will need to be bred that have thick, tender and juicy ribs. Wide collections of material must be made, especially of the old cultigens which can still be recovered, so as to characterize and select them. The areas of greatest interest are the Maghreb, southern Greece and non-horticultural Spanish regions.

If the intention is to use the roots, harvesting should be carried out until the end of the winter. Resistance to flowering will enable root yield to be improved by encouraging rapid root growth at the time of hot weather.

Tropical and Subtropical Fruit

Tropical and subtropical fruits, in contrast with temperate fruits, can be broadly defined as those meeting all of the following criteria: crops that have their origin and commercial growing areas (when such exist) in the tropics or subtropics, plants that are evergreen and perennial, crops with a limited degree of frost resistance, and plants whose growth is practically nonexistent below 50°F (10°C) (with some exceptions according to species and individual age). A distinction between tropical and subtropical is possible if one considers that tropical species are not only sensitive to temperatures below 68°F (20°C) but indeed require a climate with average mean temperatures higher than 50°F (10°C) for the coldest month (Watson and Moncur, 1985, p. 3).

Additionally most tropicals require humid environmental conditions. Examples of truly tropical crops are traditional fruits native to Southeast Asia, like mangosteen, durian, and rambutan. A good example of a typical subtropical fruit crop is the cherimoya, which when cultivated in cold subtropical areas may suffer some foliage loss during the winter with regrowth in spring. However, some fruit crops can be cultivated equally well in either the tropics or the subtropics, of which the banana and the avocado are the most outstanding examples.

Strictly speaking, the tropics extend between the Tropics of Cancer and Capricorn, at 23° north and south of the equator.

But, agronomically speaking, these boundaries are too rigid. Not only do they contain areas, especially at higher altitudes, that do not conform to the climatic characteristics generally assigned to the tropics, but regions outside this belt have coastal areas or insular climates that may exhibit climatic conditions fitting properly in the tropics. This is the reason why some climatologists have extended the region to the thirtieth parallels (Nakasone and Paull, 1998, p.1).

In any event the main feature associated with the tropics is not so much that of heat but rather steady warm temperatures throughout the year. J. A. Samson (1986, p. 1) gave a good working definition of the tropical climate: temperature averages around 80.6°F (27°C), with the warmest month being only a few degrees higher than the coldest and temperature differences between night and day, at any given time, being greater than those between winter and summer, and, finally, little variation in day length, with the longest day being less than thirteen hours long. In comparison, the subtropics have hotter summers and cooler winters. Humidity is also generally lower. Day length differences become greater with increased latitude. The limit for the subtropics is the isotherm of 50°F (10°C) average for the coldest month (Nakasone and Paull, 1998, p. 12).

Hundreds of tropical and subtropical fruits exist, but only some fifty are well known throughout most of the world (Martin et al., 1987, p.1). These are important production crops, although a considerable gap exists between world per capita consumption (54.9 kilograms per year) and estimated consumption saturation (about 100 to 120 kilograms per year) (Jansen and Subramanian, 2000). Production and trade figures allow the division of tropicals and subtropicals into three main categories (Galán Saúco, 1996) with some overlapping.

1. Major fruits, such as banana and plantain, citrus, coconut, mango, and pineapple.
2. Minor fruits, such as abiu, atemoya, avocado, breadfruit, carambola, cashew nut, cherimoya, durian, guava, jaboticaba, jackfruit, langsat, litchi, longan, macadamia, mangosteen, papaya, passion fruit, pulusan, rambutan, sapodilla, soursop, and white sapote.

3. Wild fruits belonging to diverse botanical families. These are not cultivated commercially in any country and are much in need of characterization, conservation (both in situ, including on farm, and ex situ), selection, and breeding.

Major-category fruits are cultivated in most tropical (and subtropical) countries and are well known in both local and export-import markets. Minor fruits are not so extensively cultivated, and consumption and trade tend to be more limited, both geographically and quantitatively. However, many are of considerable economic importance in their respective regional markets, as is the case with carambola, durian, and mangosteen, which are major fruits throughout Southeast Asia (Anang and Chan, 1999).

Production of major tropical and subtropical fruits in 2000		
Fruit	World production (x 1,000 t)	Important producing countries
Orange	66,055	Brazil, United States, India, Mexico, Spain, China, Italy, Egypt, Pakistan, Greece, South Africa
Banana	58,687	Burundi, Nigeria, Costa Rica, Mexico, Colombia, Ecuador, Brazil, India, Indonesia, Philippines, Papua New Guinea, Spain
Coconut	48,375	Indonesia, Philippines, India, Sri Lanka, Brazil, Thailand, Mexico, Vietnam, Malaysia, Papua New Guinea
Plantain	30,583	Colombia, Ecuador, Peru, Venezuela, Ivory Coast, Cameroon, Sri Lanka, Myanmar
Mango	24,975	India, Indonesia, Philippines, Thailand, Mexico, Haiti, Brazil, Nigeria
Papaya	8,426	Nigeria, Mexico, Brazil, China, India, Indonesia, Thailand, Sri Lanka
Avocado	2,331	Mexico, United States, Dominican Republic, Brazil, Colombia, Chile, South Africa, Indonesia, Israel, Spain
Pineapple	13,455	Philippines, India, Indonesia, China, Brazil, United States, Mexico, Nigeria, Vietnam

Botanical Aspects

Tropical and subtropical fruits include not only woody plants, such as the mango or the orange, but also herbaceous crops like the banana and vines like the passion fruit. Most botanical families can lay claim to at least one species of tropical or subtropical fruit. Franklin Martin and colleagues (1987) list some 137 families. From the botanical point of view, a fruit is the structure developed from flowers or inflorescences. In most cases the fruit consists only of the developed ovary, but it may include other parts of the flower, such as the pedicel, sepal, or receptacle, or even a portion of the seed stalk. As with temperate crops, many different fruit types appear among the tropicals and subtropicals, from single fruits, including berries, such as the avocado or orange; drupes, such as the mango; pomes, such as the loquat; capsules, such as the durian; nutlets, such as the litchi and the longan; to compound fruits, as in the typical syncarpium of the pineapple; or even a bunch of individual berries, as in the banana. To differentiate fruit crops from perennial vegetables whose fruits are also eaten, it is necessary to keep in mind that in a horticultural sense a fruit is something that is normally eaten fresh and out of hand. A number of exceptions exist, like the breadfruit and the plantain, considered fruits by all but only palatable when cooked, as if they were vegetables. Nuts, obviously not eaten out of hand, and some tree crops whose seeds are the only part eaten, are also included among tropicals and subtropicals in most horticultural books and as such are included in this entry.

Best-known tropical and subtropical fruits and their botanical families	
Family	Common names of species
Anacardiaceae	Mango, Cashew
Annonaceae	Cherimoya, Guanábana, Custard apple
Bombacaceae	Durian
Bromeliaceae	Pineapple
Cactaceae	Pitaya
Caricaceae	Papaya
Ebenaceae	Caki
Guttifferae	Mangosteen
Lauraceae	Avocado
Malphigiaceae	Acerola
Meliaceae	Langsat or Lanson
Moraceae	Breadfruit, Jackfruit
Musaceae	Banana, Plantain
Myrtaceae	Guava
Oxalidaceae	Carambola
Palmaceae	Coconut, Date
Passifloraceae	Passion fruit, Granadilla
Proteaceae	Macadamia
Rosaceae	Loquat
Rutaceae	Orange, Grapefruit, Mandarin
Sapindaceae	Litchi, Longan, Rambutan
Sapotaceae	Chicosapote, Lucuma
Solanaceae	Sweet pepino, Lulo, Tamarillo

Areas of Origin and Spread

Although most of the continents, including the islands throughout the Pacific, have contributed tropical and subtropical fruits, most of the best-known ones came from the tropical and subtropical regions of America (for example, papaya, avocado, pineapple, guava) and Asia (for example, orange and most citrus fruits, mango, banana, litchi). Only two commercially important fruits originated in Oceania, the macadamia in Australia (specifically Queensland) and the coconut in the Pacific, the latter to the extent that its origin is considered pantropical (Martin et al., 1987, p. 47). The only important fruit native to the African continent is the date. Europe, with no tropical and limited subtropical areas, has none.

Spread to the regions surrounding their areas of origin probably began early, as soon as humans realized their value in terms of nutrition and the variety they could add to the primitive diets of the time. The potential of some species to provide not only food but also shelter or clothing (some types of banana), wood, and medicine hastened distribution.

An outstanding example is the mango. Native to the Indo-Burman region, by the end of the fourth century C.E. it had spread to all the tropical countries of Southeast Asia (Galan Sauco, 1999, p. 36). The Arabs were apparently responsible for its spread to the east coast of Africa around 700 C.E. as an adjunct to their slaving ventures. Just as Malaysians introduced the banana to Madagascar some two centuries earlier, Islamic domination brought the orange to the Mediterranean and southern Europe. Crops from the Americas are not as well documented, but archaeological findings have shown connections between the cultures of Mexico and Peru dating as far back as 1000 B.C.E. (Purseglove, 1968, p.12), giving a solid opportunity for some tropical and subtropical fruits to spread around the warmer American lands.

Soon after the European discovery of America, the Old and New Worlds rapidly exchanged crops. The sixteenth-century monk Bartolome de las Casas mentioned that orange seeds were carried from the island of La Gomera (Canary Islands, Spain) to Haiti on Christopher Columbus's second voyage in 1493 (Amador de los Rios, 1851–1855, vol. 1, p. 3). It is similarly well documented that the banana was carried to Santo Domingo from the Canary Islands in 1516 (the Canaries were a routine last port of call for European ships facing an Atlantic crossing). After Columbus's voyages, a veritable avalanche of expeditions explored all corners of the world, and where the ships went, food went also, to say nothing of tasty fruits and easily propagated species. Between 1500 and 1650 Portuguese sailors connected Brazil and the Cape of Good Hope, touching Goa, Malacca, the Moluccas, Canton, and Macao, trading from there with Japan and Formosa. The Spanish Manila galleon route dominated shipping from 1565 to 1815, plying the seas between the Philippines and Mexico. Dutch, British, and French voyagers were also important in spreading tropical fruits around the world.

No hard and fast rule explains why some fruits spread quickly throughout the world while others remain limited in scope even in the twenty-first century. Several factors may be involved, among them crop adaptability, shelf life, ease of propagation (including the capacity to survive long voyages), size of the plant, multiplicity of uses (that is, other than as fresh fruit), and taste acceptance.

The excellent taste of the pineapple, the long-lasting viability of the plant's suckers as planting material, and the rapidity with which it produces fruit all account for its prompt appearance in Europe—albeit in glasshouses—and India as early as 1548 (Nagy and Shaw, 1980, p.16; Galan Sauco, 2001). Similar considerations apply to the banana and the papaya and even to woody perennial trees like the mango or the guava, which soon spread throughout the tropics and subtropics, even though their size precluded cultivation in greenhouses outside these areas. On the other hand true tropical trees are usually demanding in climate and in some cases are difficult to propagate. The mangosteen, rambutan, and durian (this last deemed by many people to have a peculiar taste) have remained confined almost exclusively to their area of origin in Southeast Asia. The mangosteen is notable among tropical fruits in that it has proven particularly intractable to most attempts to establish it outside of its area of origin via the usual method, which is selection or breeding of cultivars capable of adapting to environments different in climate or edaphic conditions. The species consists of a single genotype, which in essence means no genetic variation exists with which to breed or improve stock, and it is entirely possible that its evolution has ceased (Yaacob and Tindall, 1995, p. 25).

Nutritional and Medicinal Value

Despite the relatively low caloric values of tropical and subtropical fruits (banana and plantain and avocado are the notable exceptions), they play an important role in human diet mainly because of their high and diverse vitamin and mineral content. This has been of capital importance in the tropics, where people have been consuming them since ancient times, either by collecting fruit from the wild or by cultivating plants in kitchen gardens. They have become an important part of the diet of people in the developed countries of the world, especially among the health and fitness conscious. In a properly balanced diet, tropical and subtropical fruits may be an excellent component for the sports-oriented person. This is not to say that one can live by tropical fruits alone or that they can be considered staple fruits within the diet (again the banana and especially its relative the plantain are the exception in some tropical areas). But nutritionists have long recommended a minimum of one hundred grams of fruit per day and that it be as varied as possible. Toward the end of the twentieth century market campaigns commonly recommended consumption of five fruits per day, which, while it may have more to do with commerce than with science, does reinforce the value of fruit as a part of the human diet.

Tropical and subtropical fruits also have some medicinal properties. Many tropical fruits, notably the mango and the papaya, are a good source of carotene (provitamin A). An indication of the high content of this vitamin is the orange-yellow colour of the flesh. Others, like all citrus fruits and the guava, are well known as good sources of ascorbic acid (vitamin C). In general they are not a good source of the B group of vitamins (thiamine, riboflavin, and niacin) except for nuts, which are also a good source of vitamin E, proteins, and fats (Martin et al., 1987, p. 7). Tropical and subtropical fruits are also rich in pectin, fibre, and cellulase, which promote intestinal motility. In common with other fruits, they are good sources of antioxidants, and some are also good sources of organic acids, which stimulate appetite and aid digestion.

Values for the chemical composition of tropical and subtropical fruits are widely available in many texts, some of which are included in the bibliography cited here, but the salient points related to general nutritional value follow. Banana is a good source of vitamins A, B, and C and riboflavin. Together with the tropical and subtropical nut fruits, the banana has the highest calorie content. It is low in protein and fat and rich in potassium. Easy to digest, it constitutes an excellent food for young and old alike and is recommended for athletes. Avocado has a good oil content (of the different avocado races, the West Indian types have the lowest) composed of highly digestible unsaturated fatty acids, and it is rich in folic acid. Some cultivars contain good quantities of proteins, vitamin A, riboflavin, and phosphorus.

All citrus fruits have fairly high amounts of vitamin C, as does the guava, which in turn contains fair amounts of niacin and iron. The papaya has high quantities of vitamins C and A as well as potassium and calcium, and it is low in carbohydrates. However, its outstanding feature, which distinguishes the papaya from all other fruits, is the fact that it contains papain, an enzyme that promotes digestion (although papain content does decrease as the fruit ripens). It is highly recommended for people with certain digestive disorders. The mango is rich in provitamin A and carbohydrates and is an acceptable source of vitamin C. The same is true of the passion fruit, which additionally has acceptable quantities of niacin. The pineapple is also rich in vitamin C and carbohydrates and is a good source of calcium, phosphorus, ircn, potassium, and thiamine.

The litchi and the longan, most of the Annonaceae, and the durian are all good sources of carbohydrates and vitamin C. The durian also has fair amounts of iron and niacin. The mangosteen is considered by many to be one of the finest tasting fruits of all, according it the title of "queen of fruits" (Yaacob and Tindall, 1995, p. v). It is one of the lowest in nutritive value, but even so it can boast moderate quantities of calcium, phosphorus, ascorbic acid, and carbohydrates. The carambola is low in calories and rich in vitamin C, and it is an adequate source of vitamin A. It is prohibited for people with kidney problems (specifically stone formation) due to its high oxalic acid content, but new cultivars have been selected for lower oxalic content while maintaining sugar and vitamin levels (Galan Sauco et al., 1993, p. 5).

The macadamia nut is rich in protein, oil, iron, calcium, thiamine, riboflavin, and niacin. The subtropical date also has a high nutritive value. Rich in carbohydrates, it is a good source of vitamin A, potassium, and iron but is low in oils and sodium. The coconut is high in phosphorus, iron, proteins, and oils—in this case all saturated fatty acids, the consumption of which should be limited according to health recommendations. Coconut milk aids in balancing pH in the body due to its alkaline reaction.

The medicinal value of tropicals and subtropicals, both the fruits themselves and their actual plant parts (bark, roots, and even pollen), has long been acknowledged by the diverse peoples in and around their areas of origin. These regions are rich in recipes for preparing infusions, decoctions, syrups, pastes, jellies, juices, and so forth for myriad purposes. All the citrus fruits and several others rich in vitamin C are obviously useful to prevent colds and similar infections, while fruits rich in vitamin A prevent dietary deficiencies, such as those leading to blindness. An excellent compilation of popular medicinal uses is in the book *Fruits of Warm Climates* (1987), written by Julia F. Morton, but a few examples follow.

The date has a high tannin content that is reportedly useful as an astringent in intestinal complaints and is good for sore throats, colds, and bronchial catarrh. Breadfruit is reported to reduce high blood pressure. Carambola fruit and pineapple juice are reportedly useful diuretics, while the flesh of the very young fruit of the pineapple is reputedly an abortifacient. The skin of the avocado and extracts of ripe and unripe fruits and seeds of the papaya reportedly have antibiotic properties. In traditional medicine a decoction of young mango leaves is recommended as a remedy for asthma, blenorraghia, and bronchitis. The roots, bark, leaves, and immature fruits of many tropical fruit crops are widely used in the tropics as astringents to stop gastroenteritis, diarrhea, and dysentery. A decoction of the boiled fruit of the sapodilla has also been reported useful in treating diarrhea. The flesh of the longan has been recommended for its febrifuge and vermifuge properties and as an antidote against some types of poisons. The infusion of passion fruit leaves, rich in the glycosid passiflorine, is reported to have sedative properties.

Consumption and Other Uses

The main method of consumption of most tropical and subtropical fruits is as fresh fruit. The breadfruit is the most important exception, as it is only eaten cooked. Nuts can be eaten directly or processed (roasted, candied, and so forth). Salads, both savoury and sweet types, are prepared with many fruits. Indeed consumption is virtually as unlimited as the chef's imagination. Jams, jellies, juices (made with fresh fruits, concentrates, or frozen pulp), sauces, ice cream and sherbets, and other desserts and diverse confectionaries are typical of the uses to which tropical and subtropical fruits are put, both industrially and domestically. Infusions as social beverages, not as medicinal remedies, are made from many different fruits.

A specific product is baby food, especially made with "healthy" fruits like the banana or the papaya, based on different kinds of puree (industrially known as aseptic, chilled aseptic, or simply chilled purees). Flour is also made from the durian and the banana. Pickles and chutneys are made from many fruits, the most famous of which is mango chutney, a staple in Indian cuisine and highly esteemed by gourmets. Dips are also popular in many countries, of which perhaps the best known is avocado-based guacamole. Guava paste or spread is consumed, usually with bread and cheese, in many countries, particularly Cuba, Brazil, and the Canary Islands.

Besides their edible and pleasant fruits, the actual plants of several tropical and subtropical fruit crops are also put to good use. Descriptions of the many properties of parts other than fruits—wood, leaves, flowers, roots, seeds—are frequently dealt with in older texts (including, among others not yet mentioned, Popenoe, 1974 [1920]; Chandler, 1958; Singh, 1960; Purseglove, 1968; Ochse et al., 1972; Coronel, 1983), but a clear dearth of in-depth studies on many of these aspects is apparent. The potential of leaves or flower extracts as biological products for use against pests and diseases is in much the same situation and is an issue relevant to organic produce, of increasing importance to concerned consumers. Some outstanding examples of alternative uses follow.

Religious Uses: Some orchards of date palms in the Mediterranean are maintained solely to supply young leaves used on Palm Sunday during the Christian Easter week.

Oils, Perfumes, and the Like: An essential oil is extracted from some citrus species, particularly from certain oranges and their flowers. Avocado oil, occasionally used for cooking, is a commercial product in some countries. Soaps, bath gels, and shampoos include extracts from different tropical and subtropical fruits. Loquat seed oil is used in soaps and paints.

Animal Feed: Banana leaves, pseudostems, and fruits are fed to goats in several countries, particularly in the Canary Islands (Galan Sauco, 2001). Dried dates and their pits, breadfruit leaves, and mango seed kernels are used as feed in several countries. In India, Gandhi recommended using peanuts and mango seed kernels rather than expensive cereals and imported fodders (Galan Sauco, 1999, p. 44).

Textiles and Paper: Fibres from pineapple and banana leaves are used in several places for papermaking and cloth, notably in the Philippines to make the typical loose-fitting shirts called guayaberas.

Handicrafts: Mature date palm leaves and avocado wood are excellent for decorative carvings.

Construction and Furniture: The wood of breadfruit, citrus in general, guava, longan, mango, and mangosteen are regularly used for interior paneling or for furniture. The wood of the caki is highly prized. Banana and date palm leaves are a traditional roofing material in many regions.

Firewood: Orange wood is long lasting, while avocado wood is highly combustible. Mango wood is held in high esteem in Bangladesh, to the extent that the locals consider the best trees those that faithfully provide both wood and fruit (Galan Sauco, 1999, p. 44).

Other Uses: For many years chewing gum (chicle) was made from sapodilla latex. Although the industry subsequently began to use artificial substances, the trend in favour of organic products may signify a return to traditional chicle. Garden brooms are made out of the stripped fruit clusters of the date palm. Fishermen in the Pacific have used the coconut as a fishing aid, chewing the coconut meat and spitting the resulting mass onto the water to produce a glossy calm spot, smooth enough to allow a brief glimpse of the fish below the surface (Hawaii).

The potential for development of tropical fruits does not rely only on consumption. Planting tropical fruits for agro-forestry and for urban horticulture are important endeavours. In fact tropical countries like Malaysia encourage and promote intercropping of suitable perennial fruits with compatible forest species (Anang and Chan, 1999). Many tropical fruit trees make beautiful ornamental plants not only capable of improving air quality but also capable of contributing to ecological stability. They are easy to handle in gardens or in industrial or community buildings and are adequate for planting along country roads.

These considerations may involve new lines of research, particularly searching for cultivars that can be oriented toward wood (or flower) production. As indicated at the World Conference on Horticultural Research (WCHR) held in Rome in June 1998, international agencies and local authorities should work together with university and government scientists to promote the utilization of horticultural plants in large metropolitan areas (Gosselin et al., 1999).

Commercialization and Trade

In addition to citrus and the banana, four other tropical and subtropical fruits, pineapple, mango, avocado, and papaya, dominate the fresh fruit export trade. Pineapple clearly leads the ranking in processed fruits with a wide range of products, although juice and rings in syrup are the best known.

Many other tropical and subtropical fruits are no longer exotic products in world markets, having become firmly established with guaranteed supply and reasonable prices. Carambola, guava, litchi, mangosteen, passion fruit, and rambutan have experienced notable development. The main importers of most of these tropical and subtropical fruits are the European Union, the United States, Japan, Canada, and China.

Exports of fresh fruits are mainly by ship or surface transport. Post-harvest techniques for extending the shelf life of most tropical and subtropical fruits have been mastered, and refrigerated boats (some even providing controlled atmosphere installations) move these commodities from production countries to their ultimate markets with ease. A small proportion of the major fruits, particularly pineapple, mango, and papaya, are transported by air, either destined specially for gourmet or niche markets or for celebrations at certain times of the year, such as Christmas and New Year's, when they command higher prices. Some of the minor crops, still considered exctics, like the mangosteen and the rambutan, have a more difficult Post-harvest life and therefore are exported by air.

Many countries from virtually all the continents have designated specific areas for production of fruits destined purely for export. Those countries include India, Malaysia, Thailand, and China in Asia; the Philippines and Australia in Oceania; South Africa and Ivory Coast in Africa; Mexico, Brazil, the United States, Peru, Costa Rica, and Chile in North and South America; Spain in Europe; and Israel.

While banana, pineapple, and citrus have a long history of international trade, the avocado trade burst upon the scene in the 1970s. The mango did not become a well-known fruit (from a consumption point of view) until the 1990s, with Mexico as the leading exporter. The papaya and the litchi may still revolutionize trade.

Of particular relevance for the development of tropical and subtropical fruit trade is the World Trade Organization (WTO) agreement in Marrakech on 15 April 1994 following the conclusion of the Uruguayan round of General Agreement on Tariffs and Trade (GATT) talks. Basically these agreements established the principle of free trade not exposed to arbitrary market entrance taxes, and obligate signatory countries (in practice most of the world) to use only sanitary and phytosanitary quarantine measures based on solid scientific information, thus effectively halting the use of these measures as a loophole to arbitrarily restrict imports.

As in other commodities, an interesting market is developing for organically produced tropical and subtropical fruits, and organic pineapples and bananas are available in Western markets.

International Forum on Tropical and Subtropical Fruits

Many organizations and horticultural societies at national and international levels are dedicated to particular tropical or subtropical fruits (or a closely related group). Their members include amateurs, growers, researchers and academics, handlers, traders, and consumers. By reason of both magnitude and global concern, some of these merit special mention.

The International Society of Horticultural Science (ISHS), headquartered in Louvain, Belgium, has established a Commission of Tropical and Subtropical Horticulture with working groups in specific tropical and subtropical fruits. The ISHS meets regularly in different countries to discuss aspects of production, research, and trade of these fruits, and it holds an international congress every four years, which congregates a minimum of four thousand people.

The Interamerican Society of Tropical Horticulture was formerly known as the Tropical Region of the American Society of Horticultural Science. It holds annual meetings in different American countries with tropical crops to discuss the same issues mentioned above but including vegetables and ornamental plants. The Intergovernmental Group on Bananas and on Tropical Fruits, under the auspices of the Food and Agriculture Organization of the United Nations (FAO), meets every two years to discuss issues related to marketing and trade.

Dates

The date palm (*Phoenix dactylifera*) has been cultivated in the Middle East since ancient times, where it has assumed a role as more than simply a source of food and become culturally associated with Islamic culture. In the words of the Prophet Muhammad, "There is among trees one tree which is blessed... it is the palm."

The date palm is adapted to areas with long, very hot summers with little rain, low humidity, and abundant underground water. This is expressed by the saying that the date palm "must have its feet in running water and its head in the fire of the sky." These conditions are found in oases and river valleys in the arid subtropical deserts of the Middle East, the area of origin of the date palm. This is the "Fertile Crescent," where agriculture in the Old World is thought to have arisen. The date palm has been cultivated in this area since about 7000 B.C.E., and was possibly one of the first crops domesticated. By 2000 B.C.E., date palm culture had spread to Palestine, Arabia, Egypt, North Africa, and western India.

Date palms or their wild progenitors were undoubtedly used by man even before actual cultivation began. A date palm oasis must have been a welcome sight to those crossing the desert. Here were water, shade, and fresh and dried fruits high in carbohydrates. The dried fruits were easily stored and transported after leaving the oasis. The date palm also supplied building material, fibre, fuel, animal feed, honey (syrup), and wine.

The date palm had great spiritual and cultural significance to peoples of the region. It is depicted on many ancient tablets, bas-reliefs, and so forth. The date palm is mentioned a number of times in Jewish and Christian writings, but achieved its greatest esteem in Islamic culture. The date palm was consecrated by Muhammad in both his public and private life, and is prominently mentioned in the Koran and in other Islamic writings. Date consumption spread from Arabia along with Islam, and dates are now eaten by Muslims in areas unsuitable for their production, such as Indonesia and Thailand. Date culture eventually spread to non-Islamic countries with suitable growing conditions, but its culture and consumption in these areas is minor compared to that in the Islamic world.

In the early twenty-first century, the Middle East is still the centre of date production and consumption. The largest producers of dates are Egypt, Iran, Iraq, and Saudi Arabia. Most dates are consumed locally, but there is some export, mostly to other Islamic countries that do not have suitable growing conditions. Production of dates is highly specialized and labour-intensive. There are great variations in date growing practices: from traditional oasis culture to modern industrial plantings. The United States has led the way in mechanization of date production, but this practice is spreading to other countries as they modernize.

There are thousands of local varieties of dates grown in the Middle East. Other countries have a more limited number of varieties derived from a few importations. Recently, *barhee* and *medjool* have become increasingly prominent due to their use as foundation materials for tissue-cultured plants. The use of tissue-cultured plants has become common in some countries as the increase in land area devoted to date culture has expanded beyond that which can be planted with offshoots, the traditional method of propagation.

Dates are consumed fresh or in processed form. Fresh market dates are divided into dry, semidry, and soft varieties. In Middle Eastern countries, they are also eaten in the early *khalal* stage. Dates are nutritious, being high in carbohydrate and fibre. In most varieties, the sugar content is mostly invert sugar (glucose and fructose), with only low levels of sucrose. Processed products are more common in the Middle East, where large amounts of dates are produced, than they are elsewhere. Processed products include sugars, pastes, flours, preserves, syrups, and fermentation products.

3

Dishes and Preparations

Jackfruit is commonly used in South and Southeast Asian cuisines. It can be eaten unripe (young) or ripe, and cooked or uncooked. The seeds can also be used in certain recipes.

Unripe (young) jackfruit is also eaten whole, cooked as a vegetable. Young jackfruit has a mild flavour and distinctive texture. The cuisines of India, Bangladesh, Sri Lanka, Indonesia, and Vietnam use cooked young jackfruit. In many cultures, jackfruit is boiled and used in curries as a food staple.

- *Kathal Subzee*: Spicy vegetable with raw jackfruit from Uttar Pradesh or Punjab, India.
- *Chakka Pradaman*: Jackfruit pudding from Kerala, India.
- *Enchorer Torkari*: Curry made from unripe jackfruit from West Bengal, India.
- *Guzo Suke*: Dry spicy dish of raw jackfruit from Mangalore, India.
- *Ghariyo*: Jackfruit sweet dish from Mangalore, India.
- *Jackfruit Pappad*: Jackfruit Pappad as a snack from Mangalore, India.
- *Chakka Varatti*: Jackfruit Jam from Kerala, India.
- *Chakka Vattal*: Jackfruit Chips from Kerala, India.
- *Panasa Koora*: Traditional Jackfruit Curry from coastal Andhra, India.
- *Gudeg*: traditional dish from Yogyakarta, Central Java, Indonesia.
- *Lodeh*: traditional Indonesian vegetable dish with coconut milk.
- *Gule Nangka*: traditional Indonesia spicy curry Indonesia.

- Gatti or Gidde in Tulu where ripe jackfruit is ground with rava to form thick paste which is put on a teak wood leaf and then cooked in steam. The gidde is ready.
- An optional ingredient in *Sayur asam* (Indonesian clear soup; the name means *tamarind vegetables*)
- Also ingredient in Indonesian traditional Minangkabau cuisine.
- Jackfruit salad: Vietnamese dish with boiled young jackfruit.
- Rice and curry in Sri Lanka
- *Fanas Poli*: Sun dried Jackfruit pulp with sugar from Konkan.

The seeds can also be eaten cooked or baked like beans. They taste similar to chestnuts.

Other preparations:

- Jackfruit chips
- Asian ice desserts (including Indonesian & Filipino)
- *Turon*, a Filipino dessert made of banana and jackfruit wrapped in an eggroll wrapper
- Sometimes an added ingredient for cassava cake
- An optional ingredient in kolak (an Indonesian mung bean and coconut based dessert).
- It is thought that jackfruit is the basis for the flavour of Juicy Fruit chewing gum.
- Jackfruit candy
- Vitamin Water sells a jackfruit - guava (b+ theanine) beverage

Acerola

Acerola (Malpighia glabra), also known as Barbados cherry or wild crapemyrtle, is a tropical fruit-bearing shrub or small tree in the family Malpighiaceae, native to the West Indies and northern South America and also cultivated in India. It grows to 3 m tall, with a dense, thorny crown. The leaves are evergreen, simple ovate-lanceolate, 5-10 cm long, with an entire margin. The flowers are produced in umbels of 2-5 together, each flower 1-1.5 cm diameter, with five pink or red petals. The fruit is bright red, 1.5-2 cm diameter, containing 2-3 hard seeds. It is juicy, often as much sour as sweet in flavour, and very high in vitamin C and other nutrients. Although resembling a cherry, it is unrelated to the true cherry (*Prunus*).

Cultivation and Uses

The fruit is edible and widely consumed in the species' native area, and is cultivated elsewhere for its high vitamin C content.

In the 1950s, a manufacturer of baby food decided that apple juice was milder for infants than orange juice. The company claimed that a drop of acerola juice in an 8 oz. can of apple juice provided the amount of vitamin C of an equal amount of orange juice.

In Puerto Rico, the acerola is so prized that custom officials exercise considerable precaution to prevent exporting of acerola cuttings.

Pear

A pear is a tree of the genus Pyrus and the juicy fruit of that tree, edible in some species. The English word *pear* is probably from Common West Germanic **pera*, probably a loanword of Vulgar Latin *pira*, the plural of *pirum*, which is itself of unknown origin. The place name *Perry* can indicate the historical presence of pear trees.

Botany

Pears are native to temperate regions of the Old World, from western Europe and north Africa east right across Asia. They are medium sized trees, reaching 10-17 m tall, often with a tall, narrow crown; a few species are shrubby. The leaves are alternately arranged, simple, 2-12 cm long, glossy green on some species, densely silvery-hairy in some others; leaf shape varies from broad oval to narrow lanceolate. Most pears are deciduous, but one or two species in southeast Asia are evergreen. Most are cold-hardy, withstanding temperatures between "25 °C and "40 °C in winter, except for the evergreen species, which only tolerate temperatures down to about "15 °C. The flowers are white, rarely tinted yellow or pink, 2-4 cm diameter, and have five petals. Like that of the related apple, the pear fruit is a pome, in most wild species 1-4 cm diameter, but in some cultivated forms up to 18 cm long and 8 cm broad; the shape varies from globose in most species, to the classic 'pear-shape' of the European Pear with an elongated basal portion and a bulbous end.

The pear is very similar to the apple in cultivation, propagation and pollination.

There are about 30 species of pears.

Selected Species

- Pyrus amygdaliformis – Almond-leafed Pear
- Pyrus austriaca – Austrian Pear
- Pyrus balansae
- Pyrus bartlett
- Pyrus betulifolia
- Pyrus bosc
- pyrus bretschneideri - Ya pear
- Pyrus calleryana – Callery Pear
- Pyrus caucasica – Caucasian Pear
- Pyrus communis – European Pear
- Pyrus cordata – Plymouth Pear
- Pyrus cossonii – Algerian Pear
- Pyrus elaeagrifolia – Oleaster-leafed Pear
- Pyrus fauriei
- Pyrus kawakamii
- Pyrus korshinskyi
- Pyrus lindleyi
- Pyrus nivalis – Snow Pear
- Pyrus pashia – Afghan Pear
- Pyrus persica
- Pyrus phaeocarpa
- Pyrus pyraster – Wild Pear
- Pyrus pyrifolia – Nashi Pear
- Pyrus regelii
- Pyrus salicifolia – Willow-leafed Pear
- Pyrus salvifolia – Sage-leafed Pear
- Pyrus serrulata
- Pyrus syriaca
- Pyrus ussuriensis – Siberian Pear, Chinese fragrant pear

Diseases

Uses

Pear trees are used as food plants by the larvae of a number of Lepidoptera species.

Only three species are important for edible fruit production, the European Pear *Pyrus communis* cultivated mainly in Europe and North America, the Chinese white pear (bai li) *Pyrus xbretschneideri*, and the Nashi Pear *Pyrus pyrifolia* (also known as Asian Pear or Apple Pear), both grown mainly in eastern Asia. There are thousands of cultivars of these three species.

Other species are used as rootstocks for European and Asian pears and as ornamental trees. The Siberian Pear, *Pyrus ussuriensis* (which produces unpalatable fruit) has been crossed with *Pyrus communis* to breed hardier pear cultivars. The Bradford Pear (*Pyrus calleryana* 'Bradford') in particular has become widespread in North America and is used only for decoration. The Willow-leafed Pear (*Pyrus salicifolia*) is grown for its attractive slender, densely silvery-hairy leaves.

Pears are consumed fresh, canned, as juice, and occasionally dried. The juice can also be used in jellies and jams, usually in combination with other fruits or berries. Fermented pear juice is called perry.

Pears are the least allergenic of all fruits. Along with lamb and soya formula, pears form part of the strictest exclusion diet for allergy sufferers.

Pear wood is one of the preferred materials in the manufacture of high-quality woodwind instruments and furniture. It is also used for wood carving, and as a firewood to produce aromatic smoke for smoking meat or tobacco.

Custard-apple

In some regions of the world, custard-apple is another name for sugar-apple, a different plant in the same genus.

The Custard-apple (Annona reticulata), known in English as bullock's heart or bull's heart, and in Hindi as ramphal or Rama's fruit, is a species of Annona, native to the tropical New World, preferring a low elevation, and a warm, humid climate. It is also known to occur in pockets on the Southern Deccan Plateau in India. It is a small deciduous or semi-evergreen tree reaching 10 m tall. The leaves are alternate, simple, oblong-lanceolate, 10-15 cm long and 5-10 cm broad. The flowers are produced in clusters, each flower 2-3 cm across, with six yellow-green petals.

The fruit is variable in shape, ranging from a symmetrical globose to heart shaped, oblong or irregular. The size ranges from 7-12 cm. When ripe, the fruit is brown or yellowish, with red highlights and a varying degree of reticulation, depending on variety. The flavour is sweet and pleasant, but inferior to that of the cherimoya or sugar-apple. The latter fruit is sometimes confused with this species.

Fatty-acid methyl ester of the seed oil meets all of the major bio-diesel requirements in the USA (ASTM D 6751-02, ASTM PS 121-99), Germany (DIN V 51606) and European Union (EN 14214).

In Britain Custard-apple refers to cherimoya (Annona cherimola).

Subtropical Fruits

Ber

BER (Zizyphus *mauritiana Lamk) is* an ancient and indigenous fruit of India, China and Malaysia region. The fruits are very nutritious and are rich in vitamin C, A & B complex. The ber is one of the most common fruit trees of India and is cultivated practically all over the country. Ber fruits can be within the reach of the poor people and hence known as poor man's fruit.

Importance: Among the fruit trees, ber cultivation requires perhaps the least inputs and care. It gives good production even without irrigation and can be grown as a rainfed crop in semi-arid and arid regions. The tree can, therefore give assured income even under marginal growing conditions and provides nutritious food at very low cost. The fruit is dried and is used as a dessert fruit. It can also be preserved as a candied fruit.

Climate: The ber is a hardy fruit and grows well all over the country under varying climatic conditions and upto an elevation of 1000 meters above sea level. For its successful cultivation, it favours a hot and dry climate. The Ber crop withstands high temperature and aridity by cessation of growth, leaf fall and dormancy phase. In high humidity conditions disease and pest problems increase especially of powdery mildew which is a serious disease.

Soil: Ber plant grows on a wide variety of soils ranging from shallow to deep and from gravelly and sandy to clayey. The ber develops a deep tap root system within a short period of its growth, and as such to adverse soil conditions. Ber can also withstand alkalinity and slightly waterlogged conditions. It can, in fact, withstand and often do better than most fruits on poor soils.

Varieties: Numerous horticultural varieties of ber are grown all over India. Some of the most popular varieties are Umran, Karaka, Gola, Seb, Chhuhara, Sanaur - 2, Ilaichi and Mehrun. In Maharashtra, cv. Umran Is being grown on commercial scale. It has excellent keeping quality and transportability but flat taste.

Propagation: The Ber was commonly propagated by seeds during earlier period. But main disadvantage of this method is of heterozygosity and variability in seedling progeny. Therefore, propagation of superior varieties by patch budding is recommended. For raising a budded plantation, it is considered best to sow ber seeds in the field itself at proper distances and use the seedlings thus raised for budding *in situ*. For early germination of ber seeds, breaking of endocarp (hard seed coat) gives quick germination.

Planting Season: Monsoon season offers best choice, for raising *in situ* plantation in Maharashtra and arid and semi-arid parts of our country, at a spacing of 6 x 6 m. Protective irrigation during summer months will ensure the survival and good growth of the plants. Deheading these seedlings to ground level in the month of May helps to give new shoots by July. Patch budding these shoots in July helps to convert it into choice variety.

In Northern India, planting is done either in February-March or July-September at a spacing of 7-8 months.

After the layout, pits of 60x60x60 cm are dug. About 100 g of 10 per cent carbaryl or Aldrex dust is sprinkled on the bottom sides of pits to prevent termites. Pits are filled with top soil mixed with 20 kg farmyard manure and 1 kg super phosphates. Treated seeds or bud grafts are planted in these pits at the onset of Monsoon.

Interculturing: Area around the young plant is kept clean by weeding and hoeing. Under irrigated conditions low growing vegetables can be grown as inter crops until the ber plants assume (4-5 years) full grown. Under rainfed conditions legume crops like Moong, Moth (horse gram) and cowpea can be grown as intercrop.

Stirring the soil under the tree canopy after rains provides better aeration around roots and helps in conserving soil moisture and weed control.

Care of young orchards:

Training: If left to itself ber trees would attain bushy, large unmanageable form but the production per unit area is quite low. In order to keep the plant in manageable shape and size, trees be trained properly during first 2-3 years to build a strong framework. By providing support with bamboo stick to the new growth of sprout from either *in situ* or transplanted seedling, vertical growth is encouraged Once the growth is about 1-1.5 meter, the terminal growth Is pinched, allowing lower buds on main stem to sprout and form main branches. After sufficient growth, these primary branches be pinched to develop secondary branches. Thus plants are trained by pruning and kept in manageable shape, size with well developed framework.

Pruning: Pruning is an essential operation in ber production as fruits are borne in the axil of leaves on the young shoots of current season. Pruning is therefore, done every year to induce maximum number of new healthy shoots which bear good quality fruits. The best time for pruning was observed to be 15th April to 15th May under prevailing climatic condition of Maharashtra.

Special Horticultural Practices:

Attempts to increase fruit set, fruit size and early/delayed maturity through application of growth regulators proved to effective. The growth regulators like GA, NAA, CCC and Ethephon were used. GA (10 ppm) and NAA (10 ppm) during fruit development and Ethephon one month before harvesting were sprayed. Application of NAA increased the yield of fruits with cost benefit ratio of 1:3:30.

Irrigation: Ber is mostly grown as a rainfed crop. Irrigation as a rule not advisable. But if available could increase yield. However, increase in yield also diminishes the quality of fruit.

Nutrition: like other fruit trees, ber also requires regular application of manures and fertilisers for good yields. Application of 30-50 kg Farm yard manure, 250 g N per tree (split in 2 equal doses) and 250 g P_2O_5 per tree + 50 g K_2O per tree (single dose) per year for full grown tree (5 years and thereafter) is recommend for better yield and quality of fruits.

Plant Protection: Ber fruitfly is one of the important pests of ber, which is widely distributed throughout India. The infested fruits turn brown, rot and smell offensively. The pest can be controlled by spraying 3 to 4 times with Carbaryl 50 WP 0.2 per cent or Dimethoate 30 EC 0.03 per cent commencing from the attainment of fruits of the pea size.

Among the diseases powdery mildew is very common on ber fruits. Small whitish spots appear on young fruits, which later enlarge and cover the entire fruit. The affected fruits either drop off or become corky, mis-shapen and under-developed. Disease can be controlled by dusting with sulphur @ 150 to 200 g /tree and subsequent 3 dustings at an interval of 15-20 days.

Alternaria leaf spot and Cercospora leaf spot also appears in the form of grey spots on leaves. Both these diseases are effectively controlled by spraying dithane M-45 (0.25%) or Foltart (0.1%) as soon as the disease appear. Subsequent 2 to 3 sprays be given at an interval of 15 to 20 days depending upon the intensity of the disease.

Harvesting and yield: In Northern India, peak period of harvesting falls between February and April. While in Maharashtra harvesting extends from November to January. The fruits are harvested in 4 or 5 pickings since all the fruits on the tree do not mature at one time. The fruit picking is done by hand using a ladder. The fruits should be harvested at proper stage of maturity. The best index of the correct picking stage is the characteristic maturity colour and softness of particular cultivar after the fruit has attained the full size. The fruit requires about 120 days to reach maturity.

Under dryland (rainfall) conditions, on an average 60-80 kg fruits per tree per year can be harvested. Under irrigated conditions yield will be 3-4 times higher.

Post harvest handling and marketing: The underripe, overripe and damaged fruits are sorted out. The remaining sound fruits are graded in two grades -large and small according to size. Fruits are packed in gunny bags, wooden boxes, cardboard boxes or nylon knot bags.

The growers generally auction their crop to the contractors. The contracts are fixed either on the share of accruing incoming basis or on the basis of a lump sum to be paid in installments to the grower.

Grapes

Introduction: The grape is the most important crop grown in the world. Mostly it grown for making wines and preparation of raisin and then as a table fresh fruit. While in India, it is mainly grown for table use. Grape cultivation is believed to have originated near Caspian Sea, however, Indians know grapes since Roman times. Total area under grapes in India is about 40,000 ha, distributed mainly in Maharashtra, Karnataka, Andhra Pradesh and Tamil Nadu.

Economic Importance: At present, grape is the most important fruit crop grown commercially with the objectives.

a. For table purpose
b. For export purpose
c. For making wines and
d. For making raisins.

Fresh grapes are a fairly good source of minerals like calcium, phosphorous, iron and vitamins like B. Famous champagne and other desert wines are prepared from grapes.

Climatic Requirements: The ideal climate for grape growing is the Mediterranean climate. In its natural habitat, the vines grow and produce during the hot and dry period. Under South Indian conditions – vines produce vegetative growth during the period from April to September and then fruiting period from October to March. Temperatures above 10^{0}C to 40^{0}C influence the yield and quality. High humidity and cloudy weather invite many fungal diseases, besides lowering the T.S.S.: Acid ratio.

Soil Preference: The grape is widely adopted to various soil conditions, but the yield and quality reach to the highest on good fertile soils have pH 6.5 to 8.5, organic carbon above 1.0%, free of lime and having a medium water holding capacity. Early but medium yields with high T.S.S. are harvested on medium type of soils.

Varieties:

a. Table purpose varieties-
   i. Seeded varieties –Cardinal, concord Emperor, Italia, Anab-e-shahi, Cheema Sahebi, Kalisahebi, Rao Sahebi,
   ii. Seedless varieties – Thompson seedless, flame seedless, kishmish chorni, perlette, Arkavati.
b. Raisin purpose varieties – Thompson seedless, manik chaman, sonaka, Black corinth, Black monukka, Arkavati, Dattier
c. Wine varieties – Chardonnay, Cabernet Saurignnon, Bangalore Blue, Muscat, Blanc, Pinot Noir, Pinot Blane, White Riesling, and Merlot.

Propagation: Grapevine is most commonly propagated by hard-wood cuttings, though propagation by seed soft wood cutting, layering, grafting and budding is specific to certain situations. Occasionally, unrooted cuttings are also planted directly in the field in the pre-determined position for a vine. For hardwood cuttings, IBA, 1000 ppm treatment is useful for early, better and uniform rooting of cutting. For grafting Dog ridge, Ramsey, 1616, 1613, 1103P, So4, etc. are used. Sometimes the rootstocks are planted in the field and there they are grafted with suitable varieties.

Planting and Season: Usually planting is done from October onwards till January. Rarely planting is also done during June-July where the monsoon is late. Monsoon planting is avoided mainly for avoiding diseases on young growth. For planting N-S direction the trenches are opened. The size of trench may be 60 to 75 cm. Deep wide.

Then these trenches are filled with FYM, organic manures, 5:10:5 organic mixtures, single super phosphates, bio-fertilisers, neem cakes, etc. Spacing for planting is maintained depending on soil type, variety and method of training. The distance between two rows may be 2 to 3 m while distance between vines within a row will be half of that, accommodating vines from 2000 to 5000 per hectare.

Interculturing: The following aspects are important:-

i. Gap filling: To be done preferably during one month after planting.
ii. Recut: Basal cut keeping 2/3 buds is taken one month after planting with an objective to get uniform new growth.
iii. Supporting: The bamboo supports are fixed for vine support and young growing points are trained on them
iv. Weeding: The vine rows are weeded out, twice/thrice depending on the intênsity of weeds.
v. Irrigations – regular depending on soil and season are given.
vi. Fertilizers are applied with cow during slurry to hasten the growth.
vii. Suitable plant protection measures are following depending on incidence of pests and diseases.

Care of Young Orchard: Grapes vines takes about 1.5 to 2 years after planting to bear the first crop. During this period the care of young vines is taken as under:-

i. Training: The vines are trained first on bamboo and then on support – trellis. A suitable method of training is adopted.
ii. Pruning – Initial pruning is done only for training *i.e.* for developing trunk, arm, fruiting, canes, etc.
iii. The fertilizer doses – including organic, inorganic and bio-fertilizers are applied twice in a year.
iv. Plant protection schedule is prepared and followed for the total initial period of growing.

Special Horticultural Practices:

i. Pruning and training: The vines are trained on a suitable trellis *i.e.* 'T', 'Y', 'H' or bower and regularly pruned twice in a year. First annual pruning is done during the month of April to get the new vegetative growth while second pruning to get the crop is done during the month of

October. While doing April pruning 0 to 2 buds on arm are kept while doing October pruning 5 to 10 buds on fruiting cane are kept. Use of HCN is done to have early, uniform and higher sprouting particularly after winter pruning is made.

ii. Girdling: Vines are trunk girdled at bloom period to increase the fruit set, to increase the weight and T.S.S. and also to enhance maturity.

iii. Use of hormones: The following plant hormones at various stages and concentrations are usually used to increase the yield and to improve the quality of bunches.

A. At bloom.............GA3.. ...20 to 30 ppm; CCC ...500 ppm

B. At setting.........GA3 ....30 to 40 ppm; 6BA..... 5-10 ppm

C. At 4-6 mm size....GA3...40 to 50 ppm; Brassinos...100 ppm

D. At 6-8 mm size...GA3...30 to 50 ppm; Cppu...2-3 ppm

Irrigation: Grape is strictly irrigated perennial crop and regularly irrigated. For flood irrigation, 5-7 days during summer 8-10 days during winter and 15-20 days during rainy season – interval is maintained while for drip irrigation, 40-50 L; 30-40, 20-30 L of water per day per vine, water is applied.

Nutrition: Balanced nutrition and use of chemical, organic and bio-fertilisers is essential to get a good crop of good quality every year. About 700 to 900 N, 400 to 600 P and 750 to 1000 K Kgs/ ha/year are applied to get about 30 to 35 tonnes produced yearly.

The use of vermiphos, biomeal, mixtures of 5:10:5 ormichem, micronutrient mixtures have proved useful in grape production. Fertilizers are applied mainly twice in a year at the time of pruning, besides occasional foliar sprays are also practiced. Now-a days, Fertigation techniques is being popular in grape growers.

Plant Protection: Grape shoots, leaves, blossoms and berries are attacked by many fungi and insect pests, besides some nematodes are also cause damage to roots.

Major fungal diseases – Anthracnose, Powdery mildew, Downy mildew, Dead arm, Botraytis and Botrodiplodia.

Major bacterial diseases –Xyantomonas, blight.

Viral diseases – Fan leaf disease

Major Insect pests – Flea beetle, Mealy bug, Red mite, Thrips, caterpillars.

Soil born pests – Nematodes, phyloxere, white ants, and white grub.

Control Measures: A number of systemic and contact fungicides and pesticides are available and are to be used as per following local schedule.

Besides pests and diseases the crop is to be protected against weeds, Cyprus, doob grass, Parthenium olerace are some of the common and important weeds found in vineyards. They are controlled by frequent weeding fruit ching/growing cover crops or by using chemical weedicides as Gramaxone, Basaline, Roundup, Glycel, etc.

Grape bunches are also to be protected against hot sun, Cold wave, dry air spell, Dew and Storm. Some chemical some physiological and some mechanical means methods are adopted.

Harvesting and Yields: Normal grape harvest season starts in February and continuous up to end of April. Well matured bunches having at least 18^0 Brix are harvested

Av. yields - For seedless varieties - 20 to 30 t/ha/y

For seeded varieties - 40 to 50 t/ha/y

Post Harvest Handling: Harvested grapes are packed in 2 to 4 kg-corrugated boxes. Grape guards, pouches are kept inside the boxes for distant markets. Pre-cooling and use of grape guards are the musts for cold storage and export markets.

Mumbai, Delhi, Calcutta, Ahmedabad, Ludhiana, Patna, Jamshedpur, Bangalore, Hyderabad, are the main market places in the country.

Some additional features:

a. Export: Day by day increased quantities are exported to Europe, Middle East, Dubai, etc.

b. Winemaking: Wine making and champagne making are profitable and started at few places in Maharashtra and Karnataka.

c. Raisin making: A good quality black and golden raisins are prepared from the varieties. The raisins are increasing demand from Indian as well as from outside countries.

Olive

The Olive (*Olea europaea*) is a species of small tree in the family Oleaceae, native to coastal areas of the eastern Mediterranean region, from Lebanon and the maritime parts of Asia Minor and northern Iran at the south end of the Caspian Sea. Its fruit, the olive, is of major agricultural importance in the Mediterranean region as the source of olive oil.

Description: The Olive is an evergreen tree or shrub native to the Mediterranean, Asia and parts of Africa. It is short and squat, and rarely exceeds 8-15 meters in height.

The silvery green leaves are oblong in shape, measuring 4-10 cm long and 1-3 cm wide. The trunk is typically gnarled and twisted.

The small white flowers, with four-cleft calyx and corolla, two stamens and bifid stigma, are borne generally on the last year's wood, in racemes springing from the axils of the leaves.

The fruit is a small drupe 1-2.5 cm long, thinner-fleshed and smaller in wild plants than in orchard cultivars. Olives are harvested at the green stage or left to ripen to a rich purple colour (black olive). Canned black olives may contain chemicals that turn them black artificially.

History: The olive is one of the earliest plants cited in recorded literature. In Homer's Odyssey, Odysseus crawls beneath two shoots of olive that grow from a single stock. Horace mentions it in reference to his own diet, which he describes as very simple: "As for me, olives, endives, and smooth mallows provide sustenance." Lord Monboddo comments on the olive in 1779 as one of the foods preferred by the ancients and as one of the most perfect foods.

Old Trees

Pliny the Elder told of a sacred Greek olive tree that was 1600 years old. Several trees in the Garden of Gethsemane (from the Hebrew words "gat shemanim" or oil press) in Jerusalem are claimed to date back to the time of Jesus. Some Italian olive trees are believed to date back to Roman times. Dating and identifying the trees that appear in ancient sources is not a simple task.

However, the age of an olive tree in Crete, claimed to be over 2,000 years old, has been determined on the basis of tree ring analysis. Another, on the island of Brijuni (Brioni), Istria in Croatia, a well-known olive tree has been calculated to be about 1,600 years old. It still gives fruit (about 30kg per year), which is made into top quality olive oil. The olive tree is one of the symbols of Athena, the Greek goddess, and is frequently mentioned in the Bible and the Quran.

Cultivation and Uses

The olive has been cultivated since ancient times as a source of olive oil, fine wood and olives for consumption. The fruit, being bitter in its natural state, is typically subjected to fermentation or cured with lye or brine to make it more palatable.

Green olives and black olives are soaked in a solution of sodium hydroxide and then washed thoroughly in water to remove oleuropein, a naturally bitter carbohydrate.

Green olives are allowed to ferment before being packed in a brine solution. Black olives are not fermented, which is why they taste milder than green olives.

It is not known when olives were first cultivated for harvest. Among the earliest evidence for the domestication of olives comes from the Chalcolithic Period archaeological site of Teleilat Ghassul in what is today modern Jordan.

The plant and its products are frequently referred to in the Bible and by the earliest recorded poets. Farmers in ancient times believed that olive trees would not grow well if planted more than a short distance from the sea; Theophrastus gives 300 stadia (55.6 km) as the limit. Modern experience does not always confirm this, and, though showing a preference for the coast, it has long been grown further inland in some areas with suitable climates, particularly in the southwestern Mediterranean (Iberia, northwest Africa) where winters are less severe.

Olives are now cultivated in many regions of the world such as South Africa, Australia, New Zealand, Mediterranean Basin and California. Considerable research has been done to support the health benefits of eating olives and olive oil.

The olive tree provides leaves, fruit and oil. Olive leaves are used in medicinal teas.

Subspecies

There are at least five natural subspecies distributed over a wide range:

- *Olea europaea* subsp. *europaea* (Europe)
- *Olea europaea* subsp. *cuspidata* (Iran to China)
- *Olea europaea* subsp. *guanchica* (Canaries)
- *Olea europaea* subsp. *maroccana* (Morocco)
- *Olea europaea* subsp. *laperrinei* (Algeria, Sudan, Niger)

Cultivars

There are thousands of cultivars of the olive. In Italy alone at least three hundred cultivars have been enumerated, but only a few are grown to a large extent. The main Italian cultivars are 'Leccino', 'Frantoio' and 'Carolea'. None of these can be safely identified with ancient descriptions, though it is not unlikely that some of the narrow-leaved cultivars that are most esteemed may be descendants of the Licinian olive. The Iberian olives are usually cured and eaten, often after being pitted, stuffed (with pickled pimento, onion, or other garnishes) and jarred in fresh brine.

Since many cultivars are self sterile or nearly so, they are generally planted in pairs with a single primary cultivar and a secondary cultivar selected for its ability to fertilize the primary one, for example, 'Frantoio' and 'Leccino'. In recent times, efforts have been directed at producing hybrid cultivars with qualities such as resistance to disease, quick growth and larger or more consistent crops.

Some particularly important cultivars of olive include:

- 'Frantoio' and 'Leccino'. These cultivars are the principal participants in Italian olive oils from Tuscany. Leccino has a mild sweet flavour while Frantoio is fruity with a stronger aftertaste. Due to their highly valued flavour, these cultivars have been migrated and are now grown in other countries.
- 'Arbequina' is a small, brown olive grown in Catalonia, Spain. As well as being used as a table olive, its oil is highly valued.
- 'Empeltre' is a medium sized, black olive grown in Spain. They are used both as a table olive and to produce a high quality olive oil.
- 'Kalamata' is a large, black olive, named after the city of Kalamata, Greece, used as a table olive. These olives are of a smooth and meatlike taste.
- 'Koroneiki' originates from the southern Peloponese, around Kalamata and Mani in Greece. This small olive, though difficult to cultivate, has a high oil yield and produces olive oil of exceptional quality.
- 'Picholine' originated in the south of France. It is green, medium size, and elongated. Their flavour is mild and nutty.
- 'Lucques' originated in the south of France (Aude departement). They are green, of a large size, and elongated. The stone has an arcuated shape. Their flavour is mild and nutty.
- 'Souri' (Syrian) originated in Lebanon and is widespread in the Levant. It has a high oil yield and exceptionally aromatic flavour.
- 'Barnea' is a modern cultivar bred in Israel to be disease resistant and to produce a generous crop. It is used both for oil and for table olives. The oil has a strong flavour with a hint of green leaf. Barnea is widely grown in Israel and in the southern hemisphere, particularly in Australia and New Zealand.

- 'Maalot' is another modern, disease-resistant, Eastern Mediterranean cultivar derived from the North African 'Chemlali' cultivar. The olive is medium sized, round, has a fruity flavour and is used almost exclusively for oil production.
- 'Mission' originated on the California Missions and is now grown throughout the state. They are black and generally used for table consumption.

Growth and Propagation

Olive trees show a marked preference for calcareous soils, flourishing best on limestone slopes and crags, and coastal climate conditions. They tolerate drought well, thanks to their sturdy and extensive root system. Olive trees can be exceptionally long-lived, up to several centuries, and can remain productive for as long, provided they are pruned correctly and regularly.

The olive tree grows very slowly, but over many years the trunk can attain a considerable diameter. A. P. de Candolle recorded one exceeding 10 m in girth. They can possibly reach great age and the trees rarely exceed 15 m in height, and are generally confined to much more limited dimensions by frequent pruning. The yellow or light greenish-brown wood is often finely veined with a darker tint; being very hard and close-grained, it is valued by woodworkers.

The olive is propagated in various ways, but cuttings or layers are generally preferred; the tree roots easily in favourable soil and throws up suckers from the stump when cut down. However, yields from trees grown from suckers or seeds are poor; it must be budded or grafted onto other specimens to do well (Lewington and Parker, 114). Branches of various thickness are cut into lengths of about 1 m and, planted deeply in manured ground, soon vegetate; shorter pieces are sometimes laid horizontally in shallow trenches, when, covered with a few centimetres of soil, they rapidly throw up sucker-like shoots. In Greece, grafting the cultivated tree on the wild form is a common practice. In Italy, embryonic buds, which form small swellings on the stems, are carefully excised and planted beneath the surface, where they grow readily, their buds soon forming a vigorous shoot. Occasionally the larger boughs are marched, and young trees thus soon obtained. The olive is also sometimes raised from seed, the oily pericarp being first softened by slight rotting, or soaking in hot water or in an alkaline solution, to facilitate germination.

Where the olive is carefully cultivated, as in Languedoc and Provence, the trees are regularly pruned. The pruning preserves the flower-bearing shoots of the preceding year, while keeping the tree low enough to allow the easy gathering of the fruit. The spaces between the trees are regularly fertilized. The crop from old trees is sometimes enormous, but they seldom bear well two years in succession, and in many instances a large harvest can only be reckoned upon every sixth or seventh season.

A calcareous soil, however dry or poor, seems best adapted to its healthy development, though the tree will grow in any light soil, and even on clay if well drained; but, as remarked by Pliny, the plant is more liable to disease on rich soils, and the oil is inferior to the produce of the poorer and more rocky ground.

Fruit Harvest and Processing

Most olives today are harvested by shaking the boughs or the whole tree. Another method involves standing on a ladder and "milking" the olives into a sack tied around the harvester's waist. Using olives found lying on the ground can result in poor quality oil. In southern Europe the olive harvest is in the winter months, continuing for several weeks, but the time varies in each country, and also with the season and the kinds cultivated.

The amount of oil contained in the fruit differs greatly in the various cultivars; the pericarp is usually 60-70% oil. Typical yields are 1.5-2.2 kg of oil per tree per year.

Olive Oil History

Homer called it "liquid gold." In ancient Greece, athletes ritually rubbed it all over their body. Its mystical glow illuminated history. Drops of it seeped into the bones of dead saints and martyrs through holes in their tombs. Olive oil has been more than mere food to the peoples of the Mediterranean: it has been medicinal, magical, an endless source of fascination and wonder and the fountain of great wealth and power.

The olive tree, symbol of abundance, glory and peace, gave its leafy branches to crown the victorious in friendly games and bloody war, and the oil of its fruit has anointed the noblest of heads throughout history. Olive crowns and olive branches, emblems of benediction and purifiation, were ritually offered to deities and powerful figures: some were even found in Tutankhamen's tomb.

Traditional Fermentation

Olives freshly picked from the tree contain phenolic compounds and a unique glycoside, oleuropein, which makes the fruit unpalatable for immediate consumption. There are many ways of processing olives for table use.

Traditional methods use the natural microflora on the fruit and procedures which select for those that bring about fermentation of the fruit. This fermentation leads to three important outcomes: the leaching out and breakdown of oleuropein and phenolic compounds; the creation of lactic acid, which is a natural preservative; and a complex of flavoursome fermentation products. The result is a product which will store with or without refrigeration.

One basic fermentation method is to get food grade containers, which may include plastic containers from companies which trade in olives and preserved vine leaves. Many bakeries also recycle food grade plastic containers which are well sized for olive fermentation; they are 10 to 20 litres in capacity. Freshly picked olives are often sold at markets in 10 kg trays. Olives should be selected for their firmness if green and general good condition.

Olives can be used green, ripe green (which is a yellower shade of green, or green with hints of colour), through to full purple black ripeness. The olives are soaked in water to wash them, and drained. 7 litres (which is 7 kg) of room temperature water is added to the fermentation container, and 800 g of sea salt, and one cup (300g) of white vinegar (white wine or cider vinegar).

The salt is dissolved to create a 10% solution (the 800 g of salt is in a 8 kg mixture of salt and water and vinegar). Each olive is given a single deep slit with a small knife (if small), or up to three slits per fruit (if large, eg 60 fruit per kg). If 10 kg of olives are added to the 10% salt solution, the ultimate salinity after some weeks will be around 5 to 6% once the water in the olives moves into solution and the salt moves into the olives.

The olives are weighed down with an inert object such as a plate so they are fully immersed and lightly sealed in their container. The light sealing is to allow the gases of fermentation to escape. It is also possible to make a plastic bag partially filled with water, and lay this over the top as a venting lid which also provides a good seal. The exclusion of oxygen is useful but not as critical as when grapes are fermented to produce wine. The olives can be tasted at any time as the bitter compounds are not poisonous, and oleuropein is a useful antioxidant in the human diet.

The olives are edible within 2 weeks to a month, but can be left to cure for up to three months. Green olives will usually be firmer in texture after curing than black olives. Olives can be flavoured by soaking them in various marinades, or removing the pit and stuffing them. Herbs, spices, olive oil, feta, capsicum (pimento), chili, lemon zest, lemon juice, garlic cloves, wine, vinegar, juniper berries, and anchovies are popular flavourings. Sometimes the olives are lightly cracked with a hammer or a stone to trigger fermentation. This method of curing adds a slightly bitter taste.

Pests and Diseases

A fungus *Cycloconium oleaginum* can infect the trees for several successive seasons, causing great damage to plantations.

A species of bacterium, *Pseudomonas syringae* ssp. *savastanoi* induces tumour growth in the shoots, and certain lepidopterous caterpillars feed on the leaves and flowers, while the main damage is made by the olive-fly attacks to the fruit. In France and north-central Italy olives suffer occasionally from frost. Gales and long-continued rains during the gathering season also cause damage.

Economy

Production

Olive is the most extensively cultivated fruit crop in the world. Its cultivation areas has tripled in the past 44 years, passing from 2.6 to 8.5 million of hectares.

The first ten countries of production, all located in the Mediterranean region, represent together 95% of the world production of olives.

Main countries of production			
Year 2003	Production (in tons)	Cultivated area (in hectares)	Yield (q/Ha)
World	17 317 089	8 597 064	20,1
1. Spain	3 160 100	2 400 000	25,7
2. Italy	6 149 830	1 140 685	27,6
3. Greece	2 400 000	765 000	31,4
4. Turkey	1 800 000	594 000	30,3
5. Syria	998 988	498 981	20,0
6. Tunisia	500 000	1 500 000	3,3
7. Morocco	470 000	550 000	8,5
8. Egypt	318 339	49 888	63,8
9. Algeria	300 000	178 000	16,9
10. Portugal	280 000	430 000	6,5

Olive as an Invasive Weed

Since its first domestication, *Olea europaea* has been spreading back to the wild from planted groves. Its original wild populations in southern Europe have been largely swamped by feral plants. In some other parts of the world where it has been introduced, most notably South Australia, the Olive has become a major woody weed that displaces native vegetation.

Its seeds are spread by the introduced Red Fox and by many bird species including the European Starling and the native Emu into woodlands where they germinate and eventually form a dense canopy that prevents regeneration of native trees.

Pomegranate

The Pomegranate (*Punica granatum*) is a fruit-bearing deciduous shrub or small tree growing to 5–8 m tall. The pomegranate is native from Iran to the Himalayas in northern India and has been cultivated and naturalized over the whole Mediterranean region including Armenia since ancient times. It is widely cultivated throughout Iran, India and the drier parts of southeast Asia, Malaya, the East Indies and tropical Africa. The tree was introduced into California by Spanish settlers in 1769. In the United States, it is grown for its fruits mainly in the drier parts of California and Arizona.

Foliage and Fruit

The leaves are opposite or sub-opposite, glossy, narrow oblong, entire, 3–7 cm long and 2 cm broad. The flowers are bright red, 3 cm in diameter, with five petals (often more on cultivated plants).

The fruit is between an orange and a grapefruit in size, 7–12 cm in diameter with a rounded hexagonal shape, and has thick reddish skin and around 600 seeds. The edible parts are the seeds and the red seed pulp surrounding them. There are some cultivars which have been introduced that have a range of pulp colours like purple.

The only other species in the genus *Punica*, Socotra Pomegranate (*Punica protopunica*), is endemic to the island of Socotra. It differs in having pink (not red) flowers and smaller, less sweet fruit. Pomegranates are drought tolerant, and can be grown in dry areas with either a Mediterranean winter rainfall climate or in summer rainfall climates. In wetter areas, they are prone to root decay from fungal diseases. They are tolerant of moderate frost, down to about "10°C.

Etymology

Pomegranate, aril only

Nutritional value per 100 g (3.527 oz)

Energy 70 kcal 290 kJ

Carbohydrates	17.17 g
- Sugars 16.57 g	
- Dietary fiber 0.6 g	
Fat	0.3 g
Protein	0.95 g
Thiamin (Vit. B1) 0.030 mg	2%
Riboflavin (Vit. B2) 0.063 mg	4%
Niacin (Vit. B3) 0.300 mg	2%
Pantothenic acid (B5) 0.596 mg	12%
Vitamin B6 0.105 mg	8%
Folate (Vit. B9) 6 ?g	2%
Vitamin C 6.1 mg	10%
Calcium 3 mg	0%
Iron 0.30 mg	2%
Magnesium 3 mg	1%
Phosphorus 8 mg	1%
Potassium 259 mg	6%
Zinc 0.12 mg	1%

The name "pomegranate" derives from Latin *pomum* ("apple") and *granatus* ("seeded"). This has influenced the common name for pomegranate in many languages (*e.g.* German *Granatapfel*, seeded apple). The genus name *Punica* is named for the Phoenicians, who were active in broadening its cultivation, partly for religious reasons; consequently in classical Latin the fruit's name was *malum punicum* or *malum granatum*, where "malum" was broadly applied to many apple-like fruits. A separate, widespread root for "pomegranate" is the Egyptian and Semitic *rmn*. Attested in Ancient Egyptian, in Hebrew *rimmon*, and in Arabic *rumman*, this root was brought by Arabic to a number of languages, including Portuguese (*roma*), and Kabyle *rrumman* and Maltese "rumen".

According to the *OED*, the weapon grenade derived its name, attested in 1532, from the French name for the fruit, which is *la grenade* (from which also comes the name applied to a kind of syrup, originally pomegranate syrup, widely used in cocktails and grenadine).

Even though this fruit does not originate from China, one common nickname is "Chinese apple"

Cultivation and Uses

The pomegranate originated from Persia (Iran) and has been cultivated in Central Asia, Georgia, Armenia, Azerbaijan and the Mediterranean region for several millennia.

In Georgia, and Armenia to the east of the Black Sea, there are wild pomegranate groves outside of ancient abandoned settlements. The cultivation of the pomegranate has a long history in Armenia; decayed remains of pomegranates dating back to 1000 BC have been found in the country.

Carbonized pips and pieces of the peel of the fruit has been identified in Early Bronze Age levels of Jericho, as well as Late Bronze Age levels of Hala Sultan Tekke on Cyprus and Tiryns. A large, dry pomegranate was found in the tomb of Djehuty, the butler of Queen Hatshepsut; Mesopotamian cuneiform records mention pomegranates from the mid-Third millennium BC onwards. It is also extensively grown in South China and in Southeast Asia, whether originally spread along the route of the Silk Road or brought by sea traders.

The ancient city of Granada in Spain was renamed after the fruit during the Moorish period. Spanish colonists later introduced the fruit to the Caribbean and Latin America, but in the English colonies it was less at home: "Don't use the pomegranate inhospitably, a stranger that has come so far to pay his respects to thee" the English Quaker Peter Collinson wrote to the botanizing John Bartram in Philadelphia, 1762. "Plant it against the side of thy house, nail it close to the wall. In this manner it thrives wonderfully with us, and flowers beautifully, and bears fruit this hot year. I have twenty-four on one tree... Doctor Fothergill says, of all trees this is most salutiferous to mankind." The pomegranate had been introduced as an exotic to England the previous century, by John Tradescant the elder, but the disappointment that it did not set fruit there led to its repeated introduction to the American colonies, even New England. It succeeded in the South: Bartram received a barrel of pomegranates and oranges from a correspondent in Charleston, South Carolina, 1764. Thomas Jefferson planted pomegranates at Monticello in 1771: he had them from George Wythe of Williamsburg.

Culinary Use

After opening the pomegranate by scoring it with a knife and breaking it open, the arils (seed casings) are separated from the skin (peel) and internal white supporting structures (pith and carpellary membrane). Separating the red arils can be simplified by performing this task in a bowl of water, whereby the arils will sink and the white structures will float to the top. The entire seed is consumed raw, though the fleshy outer portion of the seed is the part that is desired. The taste differs depending on the variety of pomegranate and its state of ripeness. It can be very sweet or it can be very sour or tangy, but most fruits lie somewhere in between, which is the characteristic taste, laced with notes of its tannin.

Pomegranate juice is a popular drink in the Middle East, and is also used in Iranian and Indian cuisine; it began to be widely marketed in the US in 2004. Pomegranate concentrate is used in Syrian cuisine. Grenadine syrup is thickened and sweetened pomegranate juice; it is used in cocktail mixing. Before the tomato arrived in the Middle East, grenadine was widely used in many Persian foods; it can still be found in traditional recipes.

Wild pomegranate seeds are sometimes used as a spice, known as anardana (which literally means pomegranate (anar) seeds (dana) in Persian), most notably in Indian and Pakistani cuisine but also as a replacement for pomegranate syrup in Persian and Middle Eastern cuisine. As a result of this, the dried whole seeds can often be obtained in ethnic markets. The seeds are separated from the flesh, dried for 10–15 days and used as an acidic agent for chutney and curry production. The seeds may also be ground in order to avoid seeds becoming stuck in the teeth when eating dishes prepared with them. The seeds of the wild pomegranate *daru* from the Himalayas is considered the highest quality source for this spice.

In Turkey, Armenia and Azerbaijan, pomegranate (Turkish: *nar*; Azerbaijani: *nar*; Armenian: *nur*) is used in a variety of ways, notably as pomegranate juice. In Turkey pomegranate sauce, (Turkish: *nar ek°isi*) is used as a salad dressing, to marinate meat, or simply to drink straight. Pomegranate seeds are also used in salads, in Muhammara (Turkish Walnut Garlic Spread) and in Gullaç, a famous Turkish dessert. In Azerbaijan and Armenia, pomegranate is also used to make high-quality wine which is successfully exported to other countries.

In Greece, pomegranate is used in many recipes; such as *kollivozoumi*, a creamy broth made from boiled wheat, pomegranates and raisins; legume salad with wheat and pomegranate; traditional Middle Eastern lamb kabobs with pomegranate glaze; pomegranate eggplant relish; avocado and pomegranate dip; are just some of the dishes it is used in culinary. Pomegranate is also made into a liqueur and popular fruit confectionery that can be used as ice cream topping, or mixed with yogurt, and even spread as jams over toast for breakfast.

Health Benefits

One pomegranate delivers 40% of an adult's daily vitamin C requirement. It is also a rich source of folic acid and of antioxidants. Pomegranates are high in polyphenols. The most abundant polyphenols in pomegranate are hydrolysable tannins, particularly punicalagins, which have been shown in many peer-reviewed research publications to be the antioxidant responsible for the free-radical scavenging ability of pomegranate juice.

Many food and dietary supplement makers have found the advantages of using pomegranate extracts (which have no sugar, calories, or additives), instead of the juice, as healthy ingredients in their products. Many pomegranate extracts are essentially ellagic acid, which is largely a by-product of the juice extraction process. Ellagic acid has only been shown in published studies to absorb into the body when consumed as ellagitannins such as punicalagins.

In several human clinical trials, the juice of the pomegranate has been found effective in reducing several heart risk factors, including LDL oxidation, macrophage oxidative status, and foam cell formation, all of which are steps in atherosclerosis and heart disease. Tannins have been identified as the primary components responsible for the reduction of oxidative states which lead to these risk factors. Pomegranate has been shown to reduce systolic blood pressure by inhibiting serum angiotension converting enzyme (ACE).

Research suggests that pomegranate juice may be effective against prostate cancer and osteoarthritis. The juice can also be used as an antiseptic when applied to cuts.

Researchers at the University of Wisconsin-Madison recently discovered the potential benefits of pomegranate juice in stopping the growth of lung cancer.

Pomegranates and Symbolism

- Exodus chapter 28:33–34 directed that images of pomegranates be woven onto the borders of Hebrew priestly robes. 1 Kings chapter 7:13–22 describes pomegranates depicted in the temple King Solomon built in Jerusalem. Jewish tradition teaches that the pomegranate is a symbol for righteousness, because it is said to have 613 seeds which corresponds with the 613 mitzvot or commandments of the Torah. For this reason and others many Jews eat pomegranates on Rosh Hashanah.

- For the same reason (Exodus 28:33–34 {3}), pomegranates are a motif found in Christian religious decoration, they are often woven into the fabric on vestments and liturgical hangings, and wrought in metalwork.
- The wild pomegranate did not grow natively in the Aegean area in Neolithic times. It originated in the Iranian east and came to the Aegean world along the same cultural pathways that brought the goddess whom the Anatolians worshipped as Cybele and the Mesopotamias as Ishtar.
- The myth of Persephone, the dark goddess of the Underworld, also prominently features the pomegranate. In one version of Greek mythology, Persephone was kidnapped by Hades and taken off to live in the underworld as his wife. Her mother, Demeter (goddess of the Harvest), went into mourning for her lost daughter and thus all green things ceased to grow. Zeus, the highest ranking of the Greek gods, could not leave the Earth to die, so he commanded Hades to return Persephone. It was the rule of the Fates that anyone who consumed food or drink in the Underworld was doomed to spend eternity there. Persephone had no food, but Hades tricked her into eating four pomegranate seeds while she was still his prisoner and so, because of this, she was condemned to spend four months in the Underworld every year. During these four months, when Persephone is sitting on the throne of the Underworld next to her husband Hades, her mother Demeter mourns and no longer gives fertility to the earth. This became an ancient Greek explanation for the seasons.

It should be noted that the number of seeds that Persephone ate is varied, depending on which version of the story is told. The number of seeds she is said to have eaten ranges from three to seven, which accounts for just one barren season if it is just three or four seeds, or two barren seasons (half the year) if she ate six or seven seeds. There is no set number.

- The pomegranate also evoked the presence of the Aegean Triple Goddess who evolved into the Olympian Hera, who is sometimes represented offering the pomegranate, as in the Polykleitos' cult image of the Argive Heraion.

According to Carl A. P. Ruck and Danny Staples, the chambered pomegranate is also a surrogate for the poppy's narcotic capsule, with its comparable shape and chambered interior. On a Mycenaean seal illustrated in Joseph Campbell's *Occidental Mythology* 1964, the seated Goddess of the double-headed axe (the labrys) offers three poppy pods in her right hand and supports her breast with her left. She embodies both aspects of the dual goddess, life-giving and death-dealing at once. The Titan Orion was represented as "marrying" Side, a name that in Boeotia means "pomegranate", thus consecrating the primal hunter to the Goddess. Other Greek dialects call the pomegranate *rhoa*; its possible connection with the name of the earth goddess Rhea, inexplicable in Greek, proved suggestive for the mythographer Karl Kerenyi, who suggested that the consonance might ultimately derive from a deeper, pre-Indo-European language layer.

- In the sixth century BCE, Polykleitos took ivory and gold to sculpt the seated Argive Hera in her temple. She held a scepter in one hand and offered a pomegranate, like a royal orb, in the other. "About the pomegranate I must say nothing," whispered the traveller Pausanias in the second century AD, "for its story is something of a mystery." Indeed, in the Orion story we hear that Hera cast pomegranate-Side into dim Erebus — "for daring to rival Hera's beauty", which forms the probable point of connection with the older Osiris/Isis story. Since the ancient Egyptians identified the Orion constellation in the sky as Sah the "soul of Osiris", the identification of this section of the myth seems relatively complete. Hera wears, not a wreath nor a tiara nor a diadem, but clearly the calyx of the pomegranate that has become her serrated crown. In some artistic depictions, the pomegranate is found in the hand of Mary, mother of Jesus.
- In modern times the pomegranate still holds strong symbolic meanings for the Greeks. On important days in the Greek Orthodox calendar, such as the Presentation of the Virgin Mary and on Christmas Day, it is traditional

to have at the dinner table *"polysporia"*, also known by their ancient name "panspermia" in some regions of Greece. In ancient times they were offered to Demeter and to the other gods for fertile land, for the spirits of the dead and in honour of compassionate Dionysus. In modern times the symbolic meaning is assumed by Jesus and his mother Mary. Pomegranates are also prominent at Greek weddings and funerals. When Greeks commemorate their dead, they make *"kollyva"* as offerings that consist of boiled wheat, mixed with sugar and decorated with pomegranate. It is also traditional in Greece to break a pomegranate on the ground at weddings, on New Years and when one buys a new home for a house guest to bring as a first gift a pomegranate which is placed under/near the ikonostasi, (home altar), of the house, as it is a symbol of abundance, fertility and good luck. Pomegranate decorations for the home are very common in Greece and sold in most homegoods stores.

Other

- Pomegranate is one of the symbols of Armenia, representing fertility, abundance and marriage.
- It is the official logo of many cities in Turkey.
- The Immortals, an elite infantry unit in ancient Persia had spears with pomegranate-shaped counterweights at the butt made of gold (for officers) and silver (for regular infantry). In modern Iran the fruit is still believed to give long and healthy life.
- The Qur'an mentions pomegranates three times (6:99, 6:141, 55:068)—twice as examples of the good things God creates, once as a fruit found in the Garden of Paradise.
- Pomegranate juice stains clothing permanently unless it is washed out immediately with water — only bleach can remove stains.
- Pomegranate juice is used for natural dyeing of non-synthetic fabrics.

- Pomegranate juice is sold in the USA under several labels, and is available in health food stores and supermarkets across the country.
- Pomegranate juice will turn blue when subjected to basic (i.e. alkaline) conditions (similar to litmus paper).
- Although not native to China, Korea or Japan, the pomegranate is widely grown there and many cultivars have been developed. It is widely used for bonsai, because of its flowers and for the unusual twisted bark that older specimens can attain.
- The pomegranate also gave its name to the hand grenade from its shape and size (and the resemblance of a pomegranate's seeds to a grenade's fragments), and to the garnet from its colour. In many languages (including Belarusian, Spanish, French, Polish, and Hebrew) the words are exactly the same.
- Balaustines are the red rose-like flowers of the pomegranate, which are very bitter to the taste. In medicine, its dried form has been used as an astringent. (The term "balaustine" (Latin: *balaustinus*) is also used for a pomegranate-red colour.)
- The pomegranate was the personal emblem of the Holy Roman Emperor, Maximilian I.
- With the rise in popularity of the pomegranate in American markets, Starbucks introduced a pomegranate frappuccino in the summer of 2006.
- In the Northern hemisphere, the fruit is typically in season from September to January. In the Southern hemisphere, it is in season from March to May.
- The pomegranate is a divine symbol in Pinto Ricardo's series, The Stone Dance of the Chameleon.
- The pomegranate is one of the symbols of Hera.
- The pomegranate is also called the Food of the Dead.
- In Orthodox Christian memorial services pomegranate seeds will often be put in the koliva which is blessed after the service and eaten by all of the mourners.

Lemon

The lemon (*Citrus × limon*) is a hybrid in cultivated wild plants. The fruit are used primarily for their juice, though the pulp and rind (zest) are also used, primarily in cooking and baking. Lemon juice is about 5% acid, which gives lemons a sour taste and a pH of 2 to 3. This makes lemon juice a cheap, readily available acid for use in educational science experiments.

Description

A lemon tree can grow up to 10 meters (33 feet), but they are usually smaller. The branches are thorny, and form an open crown. The leaves are green, shiny and elliptical-acuminate. Flowers are white on the outside with a violet streaked interior and have a strong fragrance. On a lemon tree, flowers and ripe fruits can be found at the same time.

Lemon fruit are ovoid with a pointed tip at the end. When ripe, they have a bright yellow skin, a layer of pith underneath and a paler yellow segmented interior. Small seeds commonly known as 'pips' are found within the fruit.

History

The lemon is a cultivated hybrid deriving from wild species such as the citron and mandarin. When and where this first occurred is not known. The citron–apparently the fruit described in Pliny's Natural History (XII, vii.15) as the *malum medicum*, the "medicinal fruit"– seems to have been the first citrus fruit known in the Mediterranean world. Depictions of citrus trees appear in Roman mosaics of North Africa, but the first unequivocal description of the lemon is found in the early 10th-century Arabic treatise on farming by Qustus al-Rumi.

The use and cultivation of the lemon, by the Cantonese (Southern Barbarians) is noted in the early 12th century. At the end of the 12th century, Ibn Jami', personal physician to the Muslim leader Saladin, wrote a treatise on the lemon, after which it is mentioned with greater frequency in the Mediterranean. However, it is believed that the first lemons were originally cultivated in the hot, semi-arid Deccan Plateau in Central India.

The origin of the name *"lemon"* is through Persian (*Limu* [pronounced with long e and short u]), akin to the Sanskrit *nimbuka*. They were cultivated in Genoa in the mid-15th century, and appeared in the Azores in 1494. More recent research has identified lemons in the ruins of Pompeii. Lemons were once used by the British Royal Navy to combat scurvy, as they provided a large amount of Vitamin C.

In Food Preparation

Both lemons and limes are regularly served as lemonade or limeade, its equivalent, or as a garnish for drinks such as iced tea or a soft drink, with a slice either inside or on the rim of the glass. Only lemons, however, are used in the Italian liqueur Limoncello. A wedge of lemon is also often used to add flavour to water. The average lemon contains approximately 3 tablespoons of juice. Lemons warmed to room temperature before squeezing (in a microwave or by leaving on a counter) increases the amount of juice that can be extracted. Storing lemons at room temperature for long periods makes them more vulnerable to mold. Lemon juice is typically squeezed onto fish dishes; the acidic juice neutralizes the taste of amines in fish by converting them to nonvolatile ammonium salts.

In addition, lemon juice is widely used, along with other ingredients, when marinating meat before cooking: the acid provided by the juice partially hydrolyzes the tough collagen fibres in the meat (tenderizing the meat), though the juice does not have any antibiotic effects. Some people like to eat lemons as fruit; however, water should be consumed afterwards to wash the citric acid and sugar from the teeth, which might otherwise promote tooth decay and many other dental diseases. It can be used on its own or with oranges to make marmalade.

Lemons also make a good short-term preservative, commonly used on sliced apples. This keeps the fruit crisp and white for about a day, preventing the unappetizing browning effect of oxidization. This helps to prolong the usage of the fruit.

Chemistry

Lemons and other citrus fruits contain amounts of different chemicals and are thought to have some health benefits. They contain a terpene called limonene which gives their characteristic lemon smell and taste. Lemons contain significant amounts of citric acid; this is why they have a low pH and a sour taste. They also contain Vitamin C (Ascorbic acid) which is essential to human health. 100 millilitres of lemon juice contains approximately 50 milligrams of Vitamin C (55% of the recommended daily value) and 5 grams of citric acid.

Lemons can be processed to extract oils and essences.

Health Benefits

Some sources state that lemons contain unique flavonoid compounds that have antioxidant and anti-cancer properties. These may be able to deter cell growth in cancers. Limonins found in lemons could also be anti-carcinogens.

Because of its high Vitamin C content, lemon has been touted in alternative medicine as a tonic for the digestive system, immune system, and skin. There is a belief in Ayurvedic medicine that a cup of hot water with lemon juice in it tonifies and purifies the liver.

Lemon Battery

A common school experiment involving lemons is to attach electrodes and use them as a battery to power a light. The electricity generated may also be used to power a motor to move the lemons (on wheels) like a car or truck. These experiments also work with other fruit like apples and with potatoes.

Lemon Alternatives

Several other plants have a similar taste to lemons. In recent times, the Australian bush food Lemon myrtle has become a popular alternative to lemons. The crushed and dried leaves and edible essential oils have a strong, sweet lemon taste, but contain no citric acid. Lemon myrtle is popular in foods that curdle with lemon juice, such as cheesecake and ice cream.

Many other plants are noted to have a lemon-like taste or scent. Among them are Cymbopogon (lemon grass), Lemon balm, Lemon thyme, Lemon verbena, Scented geraniums, certain cultivars of basil, and certain cultivars of mint.

In popular culture:

- "The Lemon Song" by Led Zeppelin
- "Lemon" by U2
- "Life Is a Lemon and I Want My Money Back" by Meat Loaf
- "Lemon Tree" by Peter, Paul, and Mary
- "Lemon Tree" by Fool's Garden
- The word Lemon is also sometimes used to describe fanfiction that has sexually explicit scenes.

Orange (fruit)

Orange—specifically, sweet orange—refers to the citrus tree *Citrus sinensis* (syn. *Citrus aurantium* L. var. *dulcis* L., or *Citrus aurantium* Risso) and its fruit. The orange is a hybrid of ancient cultivated origin, possibly between pomelo (*Citrus maxima*) and tangerine (*Citrus reticulata*). It is a small flowering tree growing to about 10 m tall with evergreen leaves, which are arranged alternately, of ovate shape with crenulate margins and 4-10 cm long. The orange fruit is a hesperidium, a type of berry.

The word "orange" ultimately comes from Sanskrit *narang* or Tamil "naraththai". Oranges originated in southeast Asia, in either India, Vietnam or southern China. The fruit of *Citrus sinensis* is called *sweet orange* to distinguish it from *Citrus aurantium*, the bitter orange. In a number of languages, it is known as a "Chinese apple" (*e.g.* Dutch *Sinaasappel* (China's apple)).

Fruit

All citrus trees are of the single genus *Citrus*, and remain largely interbreedable; that is, there is only one "superspecies" which includes lemons, limes and oranges. Nevertheless, names have been given to the various members of the citrus family, oranges often being referred to as *Citrus sinensis* and *Citrus aurantium*. Fruits of all members of the genus *Citrus* are considered berries because they have many seeds, are fleshy, soft and derive from a single ovary. An orange seed is sometimes referred to as a pip.

Varieties

Persian Orange

The Persian orange, grown widely in southern Europe after its introduction to Italy in the 11th century, was bitter. Sweet oranges were brought to Europe in the 15th century from India by Portuguese traders, quickly displaced the bitter, and are now the most common variety of orange cultivated. The sweet orange will grow to different sizes and colours accoıding to local conditions, most commonly with ten *carpels*, or segments, inside.

Portuguese, Spanish, Arab, and Dutch sailors planted citrus trees along trade routes to prevent scurvy. On his second voyage in 1493, Christopher Columbus brought the seeds of oranges, lemons and citrons to Haiti and the Caribbean. They were introduced in Florida (along with lemons) in 1513 by Spanish explorer Juan Ponce de Leon, and were introduced to Hawaii in 1792.

Navel Orange

A single mutation in 1820 in an orchard of sweet oranges planted at a monastery in Brazil yielded the **navel orange**, also known as the Washington, Riverside or Bahia navel. The mutation causes navel oranges to develop a second orange at the base of the original fruit, opposite the stem. The second orange develops as a cojoined twin in a set of smaller segments embedded within the peel of the larger orange. From the outside, the smaller, undeveloped twin left a formation at the bottom of the fruit, looking similar to the human navel.

Because the mutation left the fruit seedless and therefore sterile, the only means available to cultivate more of this new variety is to graft cuttings onto other varieties of citrus tree. Two such cuttings of the original tree were transplanted to Riverside, California in 1870, which eventually led to worldwide popularity.

Today, navel oranges continue to be produced via cutting and grafting. This does not allow for the usual selective breeding methodologies, and so not only do the navel oranges of today have exactly the same genetic makeup as the original tree, they can even be considered to all be the fruit of that single, now centuries-old tree.

On rare occasion, however, further mutations can lead to new varieties.

Valencia Orange

The Valencia or Murcia orange is one of the sweet oranges used for juice extraction. It is a late-season fruit, and therefore a popular variety when the navel oranges are out of season. For this reason, the orange was chosen to be the official mascot of the 1982 FIFA World Cup, which was held in Spain. The mascot was called "Naranjito" ("little orange"), and wore the colours of the Spanish soccer team uniform.

Blood Orange

The blood orange has streaks of red in the fruit, and the juice is often a dark burgundy colour. The fruit has found a niche as an interesting ingredient variation on traditional Seville marmalade, with its striking red streaks and distinct flavour. The mandarin orange is similar, but smaller and sweeter, and the scarlet navel is a variety with the same diploid mutation as the navel orange.

Production

According to FAOSTAT, the top producers of oranges (in tonnes) in 2005 were:

1.	Brazil	17,804,600
2.	USA	8,393,276
3.	Mexico	4,112,711
4.	India	3,100,000
5.	China	2,412,000
6.	Spain	2,294,600
7.	Italy	2,201,025
8.	Iran	1,900,000
9.	Egypt	1,789,000
10.	Pakistan	1,579,900

Juice and other Products

Oranges are widely grown in warm climates worldwide, and the flavours of orange vary from sweet to sour. The fruit is commonly peeled and eaten fresh, or squeezed for its juice. It has a thick bitter rind that is usually discarded, but can be processed into animal feed by removing water using pressure and heat. It is also used in certain recipes as flavouring or a garnish. The outer-most layer of the rind is grated or thinly veneered with a tool called a zester, to produce orange zest, popular in cooking because it has a flavour similar to the fleshy inner part of the orange. The white part of the rind, called the pericarp or albedo and includes the pith, is a source of pectin and has nearly the same amount of vitamin C as the flesh.

Products made from oranges include:

- Orange juice, one of the commodities traded on the New York Board of Trade. Brazil is the largest producer of orange juice in the world, followed by the USA.
- Sweet orange oil, a by-product of the juice industry produced by pressing the peel. It is used as a flavouring of food and drink and for its fragrance in perfume and aromatherapy. Sweet orange oil consists of about 90% d-Limonene, a solvent used in various household chemicals, such as to condition wooden furniture, and along with other citrus oils in grease removal and as a hand-cleansing agent. It is an efficient cleaning agent which is environmentally friendly, and much less toxic than petroleum distillates. It also smells more pleasant than other cleaning agents.
- The orange blossom, which is the state flower of Florida, is traditionally associated with good fortune, and was popular in bridal bouquets and head wreaths for weddings for some time. The petals of orange blossom can also be made into a delicately citrus-scented version of rosewater. Orange blossom water is a common part of Middle Eastern cuisine. The orange blossom gives its touristic nickname to the *Costa del Azahar* ("Orange-blossom coast"), the Valencia seaboard.

- In Spain, fallen blossoms are dried and then used to make tea.
- Orange blossom honey, or actually citrus honey, is produced by putting beehives in the citrus groves during bloom, which also pollinates seeded citrus varieties. Orange blossom honey is highly prized, and tastes much like orange.
- Marmalade, a conserve made usually with Seville oranges. All parts of the orange are used to make marmalade: The pith and pips are separated, and typically placed in a muslin bag where they are boiled in the juice (and sliced peel) to extract their pectin, aiding the setting process.
- Orange peel is used by gardeners as a Slug repellent.

Since oranges are susceptible to frost damage, growers commonly use sprinklers to coat them with ice when temperatures go below freezing. This practice protects the crops by regulating temperature.

Etymology

Orange derives from Sanskrit *nraEga%* "orange tree", but another explanation tries to establish a link to a Dravidian root meaning "fragrant". Compare Tamil *narandam* "bitter orange", *nagarukam*"sweet orange" and *nari "fragrance"*. The Sanskrit or Dravidian word was borrowed into European languages through Persian *narang*, Armenian *narinj*, Arabic *naranj*, (Spanish *naranja* and Portuguese *laranja*), Late Latin *arangia*, Italian *arancia* or *arancio*, and Old French *orenge*, in chronological order. The first appearance in English dates from the 14th century. The forms starting with n- are older; this initial n- may have been mistaken as part of the indefinite article, in languages with articles ending with an -n sound (*e.g.*, in French *une norenge* may have been taken as *une orenge*). The name of the colour is derived from the fruit, first appearing in this sense in 1542.

Some languages have different words for the bitter and the sweet orange, such as Modern Greek *nerantzi* and *portokali* ("Portuguese"), respectively. The reason for these is that the sweet orange was brought from China to Europe during the

14th century by the Portuguese. For the same reason, some languages refer to it by "Applesin" which means "Apple from China".

Storage

Oranges should be stored in the warmest part of the refrigerator. They can normally be stored for about 2 weeks.

Citrus

Citrus is a common term and genus of flowering plants in the family Rutaceae, originating in tropical and subtropical southeast Asia. The plants are large shrubs or small trees, reaching 5–15 m tall, with spiny shoots and alternately arranged evergreen leaves with an entire margin. The flowers are solitary or in small corymbs, each flower 2–4 cm diameter, with five (rarely four) white petals and numerous stamens; they are often very strongly scented. The fruit is a *hesperidium*, a specialised berry, globose to elongated, 4–30 cm long and 4–20 cm diameter, with a leathery rind surrounding segments or "liths" filled with pulp vesicles. The genus is commercially important as many species are cultivated for their fruit, which is eaten fresh or pressed for juice.

Citrus fruits are notable for their fragrance, partly due to flavonoids and limonoids (which in turn are terpenes) contained in the rind, and most are juice-laden. The juice contains a high quantity of citric acid giving them their characteristic sharp flavour. They are also good sources of vitamin C and flavonoids.

The taxonomy of the genus is complex and the precise number of natural species is unclear, as many of the named species are clonally-propagated hybrids, and there is genetic evidence that even the wild, true-breeding species are of hybrid origin. Cultivated *Citrus* may be derived from as few as four ancestral species. Numerous natural and cultivated origin hybrids include commercially important fruit such as the orange, grapefruit, lemon, some limes, and some tangerines. Recent research has suggested that the closely related genus *Fortunella*, and perhaps also *Poncirus* and the Australian genera *Microcitrus* and *Eremocitrus*, should be included in *Citrus*. In fact, most botanists now classify *Microcitrus* and *Eremocitrus* as part of the genus *Citrus*.

Cultivation

As citrus trees hybridise very readily (*e.g.*, seeds grown from Persian limes can produce fruit similar to grapefruit), all commercial citrus cultivation uses trees produced by grafting the desired fruiting cultivars onto rootstocks selected for disease resistance and hardiness.

The colour of citrus fruits only develops in climates with a (diurnal) cool winter. In tropical regions with no winter, citrus fruits remain green until maturity, hence the tropical "green orange". The lime plant in particular is extremely sensitive to cool conditions, thus it is usually never exposed to cool enough conditions to develop a colour. If they are left in a cool place over winter, the fruits will actually change to a yellow colour. Many citrus fruits are picked while still green, and ripened while in transit to supermarkets.

Citrus trees are not generally frost hardy. *Citrus reticulata* tends to be the hardiest of the common Citrus species and can withstand short periods down to as cold as "10°C, but realistically temperatures not falling below "2 °C are required for successful cultivation. A few hardy hybrids can withstand temperatures well below freezing, but do not produce quality fruit. A related plant, the Trifoliate orange (*Poncirus trifoliata*) can survive below "20 °C; its fruit are astringent and inedible unless cooked.

The trees do best in a consistently sunny, humid environment with fertile soil and adequate rainfall or irrigation. (Older 'abandoned' Citrus in low valley-land may suffer, yet survive, the dry summer of Central California Inner Coast Ranges. Any age Citrus grows well with infrequent irrigation in partial/understory shade, but the fruit crop is smaller.) Though broad-leaved, they are evergreen and do not drop leaves except when stressed. The trees flower (sweet-scented at 2 to 20 meters) in the spring, and fruit is set shortly afterward. Fruit begins to ripen in fall or early winter months, depending on cultivar, and develops increasing sweetness afterward. Some cultivars of tangerines ripen by winter. Some, such as the grapefruit, may take up to eighteen months to ripen. Major commercial citrus growing areas include southern China, the Mediterranean Basin (including Southern Spain), South Africa, Australia, the southern-most United States, and parts of South America. In the U.S., Florida, Texas, and California are major producers, while smaller plantings are present in other Sun Belt states.

Citrus trees grown in tubs and wintered under cover were a feature of Renaissance gardens, once glass-making technology enabled sufficient expanses of clear glass to be produced. The *Orangerie* at the Palace of the Louvre, 1617, inspired imitations that were not eclipsed until the development of the modern greenhouse in the 1840s. An orangery was a feature of royal and aristocratic residences through the 17th and 18th centuries. In the United States the earliest surviving orangery is at the Tayloe House, Mount Airy, Virginia.

Some modern hobbyists still grow dwarf citrus in containers or greenhouses in areas where it is too cold to grow it outdoors. Consistent climate, sufficient sunlight, and proper watering are crucial if the trees are to thrive and produce fruit. Compared to many "normal green" shrubs, citrus better-tolerates poor container care. For cooler winter areas, lime and lemon should not be grown, since they are more sensitive to winter cold than other citrus fruits. Lemons are commercially grown in cooler-summer/moderate-winter coastal Southern California, because sweetness is neither attained nor expected in retail lemon fruit. Tangerines, tangors and yuzu can be grown outside even in regions with sub-zero winters, although this may affect fruit quality. Hybrids with kumquats (citrofortunella) have good cold resistance.

Pests and Diseases

Citrus plants are very liable to infestation by aphids, whitefly and scale insects (*e.g.* California red scale). Also rather important are the viral infections to which some of these ectoparasites serve as vectors such as the aphid-transmitted *Citrus tristeza virus* which when unchecked by proper methods of control is devastating to citrine plantations. The foliage is also used as a food plant by the larvae of some Lepidoptera species including Common Emerald, Double-striped Pug, Giant Leopard Moth, *Hypercompe eridanus*, *Hypercompe icasia* and *Hypercompe indecisa*. European brown snail (Helix) can be a problem in California, though laying female Mallard-based (Anas) ducks eat snails and slugs..

Uses

Culinary

Many citrus fruits, such as oranges, tangerines, grapefruits, and clementines, are generally eaten fresh. They are typically peeled and can be easily split into segments. Grapefruit is more commonly halved and eaten out of the skin with a utensil. Orange and grapefruit juices are also very popular breakfast beverages. More astringent citrus, such as lemons and limes are generally not eaten on their own. Though 'Meyer' "Lemon" can be eaten 'out of hand', it is both sweet and sour. Lemonade or limeade are popular beverages prepared by diluting the juices of these fruits and adding sugar. Lemons and limes are also used as garnishes or in cooked dishes. Their juice is used as an ingredient in a variety of dishes, it can commonly be found in salad dressings and squeezed over cooked meat or vegetables. A variety of flavours can be derived from different parts and treatments of citrus fruits. The rind and oil of the fruit is generally very bitter, especially when cooked. The fruit pulp can vary from sweet and tart to extremely sour. Marmalade, a condiment derived from cooked orange and lemon, can be especially bitter. Lemon or lime is commonly used as a garnish for water, soft drinks, or cocktails. Citrus juices, rinds, or slices are used in a variety of mixed drinks. The skin of some citrus fruits, known as zest, is used as a spice in cooking. The zest of a citrus fruit, preferably lemon or an orange, can also be soaked in water in a coffee filter, and drank.

Medical

Citrus juice also has medical uses - the lemon juice is used to relieve the pain of bee stings. The orange is also used in Vitamin C pills, which prevents scurvy. Scurvy is caused by

Vitamin C deficiency, and can be prevented by having 10 milligrams of Vitamin C a day. An early sign of scurvy is fatigue. If ignored, later symptoms are bleeding and brusing easily.

Lychee

The Lychee (*Litchi chinensis*), also spelled Litchi (the U.S. FDA spelling) or Laichi, is the sole member of the genus *Litchi* in the soapberry family Sapindaceae. It is a tropical fruit tree native from southern China. It is also found south to Indonesia and east to the Philippines. Local names include v£i, lÇ chi or (pinyin: lìzhî), Alupag (Philippines), lin jee (Thailand), Raichi (Japan) and Leechee (India).

It is a medium-sized evergreen tree, reaching 15–20 m tall, with alternate pinnate leaves, each leaf 15–25 cm long, with 2-8 lateral leaflets 5–10 cm long; the terminal leaflet is absent. The newly emerging young leaves are a bright coppery red at first, before turning green as they expand to full size. The flowers are small, greenish-white or yellowish-white, produced in panicles up to 30 cm long.

The fruit is a drupe, 3–4 cm long and 3 cm in diameter. The outside is covered by a red, roughly-textured rind that is inedible but easily removed. The inside consists of a layer of sweet, translucent white flesh, rich in vitamin C, with a texture somewhat similar to that of a grape. The edible flesh consists of a highly developed aril enveloping the seed. The centre contains a single glossy brown nut-like seed, 2 cm long and 1–1.5 cm in diameter. The seed, similar to a buckeye seed, is slightly poisonous and should not be eaten. The fruit matures from July to October, about 100 days after flowering.

There are two subspecies:

- *Litchi chinensis* subsp. *chinensis*. China, Indochina. Leaves with 4 to 8 (rarely 2) leaflets.
- *Litchi chinensis* subsp. *philippinensis* (Radlk.) Leenh. Philippines, Indonesia. Leaves with 2-4 (rarely 6) leaflets.

History

A major early Chinese historical reference to lychees was made in the Tang Dynasty, when it was the favourite fruit of

Emperor Li Longji (Xuanzong)'s favoured concubine Yang Yuhuan (Yang Guifei). The emperor had the fruit, which was only grown in southern China, delivered by the imperial messenger service's fast horses, whose riders would take shifts day and night in a Pony Express-like manner, to the capital. (Most historians believe the fruits were delivered from modern Guangdong, but some believe they came from modern Sichuan.)

The lychee was first described in the West by Pierre Sonnerat (1748–1814) on a return from his travel to China and Southeast Asia.

It was then introduced to the Reunion Island in 1764 by Joseph-Francois Charpentier de Cossigny de Palma. It was later introduced to Madagascar which has become a major producer.

There is a Cantonese saying: "*One lychee equals three torches of fire*" It refers to the extreme Yang property of the fruit.

Cultivation and Uses

Lychees are extensively grown in the native region of China, and also elsewhere in South-East Asia, India, southern Japan, and more recently in California, Hawaii, and Florida in the United States, the wetter areas of eastern Australia and sub-tropical regions of South Africa, also in the state of Sinaloa in Mexico. They require a warm subtropical to tropical climate that is cool but also frost-free or with only very slight winter frosts not below -4°C, and with high summer heat, rainfall, and humidity. Growth is best on well-drained, slightly acidic soils rich in organic matter. A wide range of cultivars is available, with early and late maturing forms suited to warmer and cooler climates respectively. They are also grown as an ornamental tree as well as for their fruit.

Lychees are commonly sold fresh in Chinese and Asian markets, and in recent years, also widely in supermarkets worldwide. The red rind turns dark brown when the fruit is refrigerated, but the taste is not affected. It is also sold canned year-round. The fruit can be dried with the rind intact, at which point the flesh shrinks and darkens, somewhat resembling a human earlobe in texture.

Cultivars

There are many different cultivars of lychee, of which three are considered to be the most sought-after.

The "Three Prestigious Cultivars":

- Hanging Green (Chinese): The most famous (and most rare) lychee in existence. It received its name because of the barely noticeable light green hue and green line on the shell. Ancient records have described Hanging Green as "Fresh and crispy as pear, without juice. It can last for three days after the shell is removed". For centuries, Hanging Green is an item of tribute to the imperial government of various dynasties, until people in Canton revolted during the Qianlong era against the tributes and chopped all but one of the Hanging Green trees. The sole remaining tree still produces fruit each year, and fruits from that tree are now called "Zhengcheng Hanging Green".
- Sweet Osmanthus Flavour: Named because of the Sweet Osmanthus flavour it contains, this lychee has light red shells, which contains sharp edges. The fruits are described as crispy and sweet. There is a related cultivar, called "Yatou Green". The shell of this cultivar has dark green spots.
- Glutinous Rice Ball: Named after its thick fruit meats and sweet (some described the taste as close to honey) flavours. The fresh red shells are not sharp and hard, and the seeds from this cultivar are noticeably smaller than others. Some fruits from this cultivar are seedless.

Other Notable Cultivars:

- Baila
- Baitangen
- Black Leaves: This cultivar matures less than others, and has big meats and seeds. The shell exhibits a dark red tint.
- Huaichi: Literally "Branches (of fruit) in the arms of (a person)", this lychee supposedly received its name when a government official toured Lingnan (mondern day

Canton) and placed within his arms lychee branches gifted by local villagers.

- March Red: This lychee matures the earliest, and are usually available annually around May.
- The Concubine Smiles: Famed as the cultivar of lychee Emperor Xuanzong of Tang brought from the edges of the Tang empire to cheer up Yang Guifei, this lychee matures earlier than others, and has a very light red tint on its shells.
- The Jade Purse: Named because of its large fruits and the thick meat within. The seed is small in this cultivar.

Nutrition

The following data was compiled by the USDA, and pertains to the nutrition information of significance per 100 grams of lychee fruit:

- calories: 66
- carbohydrates: 16.53 g
- lipids (fat): .44 g
- fibre: 1.3 g
- sugars: 15.23 g
- calcium: 5 mg
- magnesium: 10 mg
- potassium: 171 mg
- phosphorus: 31 mg
- vitamin c: 71.5 mg

Avocado

The **avocado** (*Persea americana*) is a tree native to Mexico and Central America, classified in the flowering plant family Lauraceae. The name "avocado" is also applied to the fruit of the avocado tree. The tree grows to 20 m (65 ft.), with alternately arranged, evergreen leaves, 12-25 cm long. The flowers are inconspicuous, greenish-yellow, 5-10 mm wide. The pear-shaped fruit is botanically a berry, from 7 to 20 cm long, weighs between 100 to 1000 g, and has a large central seed, 3 to 5 cm in diameter.

An average avocado tree produces about 120 avocados annually. Commercial orchards produce an average of 7 tonnes per hectare each year, with some orchards achieving 20 tonnes per hectare. Biennial bearing can be a problem, with heavy crops in one year being followed by poor yields the next. The fruit is sometimes called an avocado pear or alligator pear, due to its shape and rough green skin. The avocado tree does not tolerate freezing temperatures, and so can be grown only in subtropical and tropical climates.

Etymology

The word avocado comes from the Spanish word *aguacate*, which derives in turn from the Nahuatl (Aztec) word, *ahuacatl*, meaning "testicle", because of its shape. In some countries of South America such as Argentina, Bolivia, Chile, Peru, and Uruguay, the avocado is known by its Quechua name, *palta*. In other Spanish-speaking countries it is called *aguacate*, and in Portuguese it is *abacate*. The name "avocado pear" is sometimes used in English, as is alligator pear and "butter pear." The Nahuatl *ahuacatl* can be compounded with other words, as in *ahuacamolli*, meaning "avocado soup or sauce", from which the Mexican Spanish word *guacamole* derives.

Cultivation

The subtropical species needs a climate without frost and not too much wind. When frost does occur, the fruit drops from the tree, reducing the yield, although the cultivar 'Hass' can tolerate temperatures down to "1°C. The trees also need well aerated soils, ideally more than 1 m deep. Yield is reduced when the irrigation water is highly saline. These soil and climate conditions are met only in a few areas of the world, particularly in southern Spain, Israel, South Africa, Peru, northern Chile, Vietnam, Indonesia, Australia, New Zealand, the United States, The Philippines, Malaysia, Mexico and Central America, the centre of origin and diversity of this species. In the U.S., avocados are primarily produced in California, Florida, and Hawaii. Each region has different types of cultivars.

Propagation and Rootstocks

While an avocado propagated by seed can bear fruit, it will take 4-6 years to do so, and the offspring is unlikely to resemble the parent cultivar in fruit quality. Thus, commercial orchards are planted using grafted trees and rootstocks. Rootstocks are propagated by seed (seedling rootstocks) and also layering (clonal rootstocks). After about a year of growing the young plants in a greenhouse, they are ready to be grafted. Terminal and lateral grafting is normally used. The scion cultivar will then grow for another 6-12 months before the tree is ready to be sold. Clonal rootstocks have been selected for specific soil and disease conditions, such as poor soil aeration or resistance to the soil borne disease caused by Phytophthora root rot.

Breeding

The species is partially unable to self-pollinate, because of dichogamy in its flowering. The limitation, added to the long juvenile period, make it difficult to breed this species. Most cultivars are clonally propagated (via grafting), having originated from random seedling plants or minor mutations derived from cultivars. Modern breeding programmes tend to use isolation plots where the chances of cross-pollination are reduced. That is the case of programmes at the University of California-Riverside, as well as the Volcani Centre in Israel.

Diseases

Harvest and Post-harvest

The avocado fruit does not ripen on the tree, but will fall off or be picked in a hard, "green" state, then it will ripen quickly on the ground, but depending on the amount of oil that it has, the taste may be very different. Generally, the fruit is picked once it reaches a mature size, and will then ripen in a few days (faster if stored with other fruit such as bananas, because of the influence of ethylene gas). Premium supermarkets sell pre-softened avocados treated with synthetic ethylene to hasten the ripening process. The fruit can be left on the tree until required, rather than picked and stored, but for commercial reasons it must be picked as soon as possible. Growers can keep the fruit on the tree for about 4-6 months after fully developed; if the fruit stays on the tree for too long it will fall to the ground.

Tomatoes in Britain

The tomato plant was not grown in England until the 1590s, according to Smith. One of the earliest cultivators was John Gerard, a barber-surgeon. Gerard's *Herbal*, published in 1597 and largely plagiarized from continental sources, is also one of the earliest discussions of the tomato in England. Gerard knew that the tomato was eaten in both Spain and Italy. Nonetheless, he believed that it was poisonous (tomato leaves and stems contain poisonous glycoalkaloids, but the fruit is safe). Gerard's views were influential, and the tomato was considered unfit for eating (though not necessarily poisonous) for many years in Britain and its North American colonies. By the mid-1700s, however, tomatoes were widely eaten in Britain; and before the end of that century, the *Encyclopaedia Britannica* stated that the tomato was "in daily use" in soups, broths, and as a garnish. Tomatoes were originally known as "Love Apples", possibly based on a mistranslation of the Italian name *pomo d'oro* (golden apple) as *pomo d'amore*.

North America

The earliest reference to tomatoes in British North America is from 1710, when herbalist William Salmon reported seeing them in what is today South Carolina. They may have been introduced from the Caribbean. By the mid-18th century, they were cultivated on some Carolina plantations, and probably in other parts of the South as well. It is possible that some people continued to think tomatoes were poisonous at this time; and in general, they were grown more as ornamental plants than as food. Cultured people like Thomas Jefferson, who ate tomatoes in Paris and sent some seeds home, knew the tomato was edible, but many of the less well-educated did not.

Tomatoes in France

The tomato was introduced to France through Provence from Italy during the late 18th century and became a culinary symbol of the French Revolution due to its red colour. They are widely eaten in French cuisine. France is home to the 'Carolina', a rare, indeterminate, open-pollinated cultivar of tomato which possesses the tanginess of 'Brandywine' and the stature and externalities of the Early Swedish, that is, IPB. First noted by Italian monk Giacomo Tiramisunelli and his companion Andrea di Milininese somewhere near Bordeaux, more modern researches such as Dragos Niculae et al. and Nicolas Dela Nisan claim Belgium as the birthplace of the cultivar. Either way, the 'Carolina' is considered a rare delicacy amongst tomato-connoisseurs throughout France and beyond; it is the only cultivar of tomato traditionally served with Ortolan (fig-fed songbird). Claims that a San Diego-based U.S. biotech company is trying to genetically modify 'Carolina' to extend its potential geographic growth range has set off a minor furor in Bordeaux, with the president of a Belgian agro-commune, Victor DePlata, threatening extreme action.

Production Trends

In 2005, China was the largest producer of tomatoes as per FAO. China accounted for at least one-fourth of the global output followed by USA and Turkey.

Cultivation and Uses

The tomato is now grown worldwide for its edible fruits, with thousands of cultivars having been selected with varying fruit types, and for optimum growth in differing growing conditions. Cultivated tomatoes vary in size from cherry tomatoes, about the same 1–2 cm size as the wild tomato, up to beefsteak tomatoes 10 cm or more in diameter. The most widely grown commercial tomatoes tend to be in the 5–6 cm diameter range. Most cultivars produce red fruit; but a number of cultivars with yellow, orange, pink, purple, green, or white fruit are also available. Multicolored and striped fruit can also be quite striking. Tomatoes grown for canning are often elongated, 7–9 cm long and 4–5 cm diameter; they are known as plum tomatoes.

Tomatoes are one of the most common garden vegetables in the United States and, along with zucchini, have a reputation for outproducing the needs of the grower.

As in most sectors of agriculture, there is increasing demand in developed countries for organic tomatoes, as well as heirloom tomatoes, to make up for flavour and texture faults in commercial tomatoes. Quite a few seed merchants and banks provide a large selection of heirloom seeds. Tomato seeds are occasionally organically produced as well, but only a small percentage of organic crop acreage is grown with organic seed.

Growing Needs

Cultivars

There are a great many tomato cultivars grown for various purposes. This section attempts a listing of some of the more common cultivars. Heirloom cultivars are becoming increasingly popular, particularly among home gardeners and organic producers, since they tend to produce more interesting and flavourful crops at the possible cost of some disease resistance. Hybrid plants remain common, however, since they tend to be heavier producers and sometimes combine unusual characteristics of heirloom tomatoes with the ruggedness of conventional commercial tomatoes.

Tomato cultivars are roughly divided into several categories, based mostly on shape and size. "Slicing" or "globe" tomatoes are the usual tomatoes of commerce; beefsteak are large tomatoes often used for sandwiches and similar applications; plum tomatoes, or paste tomatoes, are bred with a higher solid content for use in tomato sauce and paste; and cherry tomatoes are small, often sweet tomatoes generally eaten whole in salads.

Tomatoes are also commonly classified as determinate or indeterminate. Determinate, or bush, types bear a full crop all at once and top off at a specific height; they are often good choices for container growing. Indeterminate cultivars develop into vines that never top off and continue producing until killed by frost. As an intermediate ground, there are plants sometimes known as "vigorous determinate" or "semi-determinate"; these top off like determinates but produce a second crop after the initial crop. Many, if not all, tomatoes described as heirlooms are indeterminate.

Commonly grown cultivars include:

- 'Beefsteak VFN' (a common hybrid resistant to Verticillium, Fusarium, and Nematodes)
- 'Big Boy' (a very common determinate garden cultivar in the United States)
- 'Black Krim' (a purple-and-red cultivar from the Crimea)
- 'Brandywine' (a pink, indeterminate beefsteak type with a considerable number of substrains)
- 'Burpee VF' (an early attempt by W. Atlee Burpee at disease resistance in a commercial tomato)
- 'Early Girl' (an early maturing globe type)
- 'Gardener's Delight' (a smaller English cultivar)
- 'Juliet' (a grape tomato developed as a substitute for the rare Santa F1)
- 'Marmande' (a heavily ridged cultivar from southern France; similar to a small beefsteak and available commercially in the U.S. as UglyRipe)
- 'Moneymaker' (an English greenhouse cultivar)
- Mortgage Lifter (a popular heirloom beefsteak known for gigantic fruit)
- 'Patio' (bred specifically for container gardens)
- 'Purple Haze' (large cherry, indeterminate. Derived from Cherokee Purple, Brandywine and Black Cherry)
- 'Roma VF' (a plum tomato common in supermarkets)
- 'Rutgers' (a commercial heirloom cultivar)
- 'San Marzano' (a plum tomato popular in Italy)
- 'Santa F1' (a Chinese grape tomato cultivar popular in the U.S. and parts of southeast Asia)
- 'Shephard's Sack' (a large variety popular in parts of Wales)
- 'Sweet 100' (a very prolific, indeterminate cherry tomato)
- 'Yellow Pear' (a yellow, pear-shaped heirloom cultivar)

Most modern tomato cultivars are smooth surfaced; but some older tomato cultivars and most modern beefsteaks often show pronounced ribbing, a feature that may have been common to virtually all pre-Columbian cultivars. In addition, some tomato cultivars produce fruit in colours other than red, including yellow, orange, pink, black, brown, and purple, though such fruit is not widely available in grocery stores, nor are their seedlings available in typical nurseries, but must be bought as seed, often via mail-order. Likewise, some less common varieties have fruit fuzzy skin, as is the case with the Fuzzy Peach tomato and Red Boar tomato plants.

There is also a considerable gap between commercial and home-gardener cultivars; home cultivars are often bred for flavour to the exclusion of all other qualities, while commercial cultivars are bred for such factors as consistent size and shape, disease and pest resistance, and suitability for mechanized picking and shipping.

Diseases and Pests

Tomato cultivars vary widely in their resistance to disease. Modern hybrids focus on improving disease resistance over the heirloom plants. One common tomato disease is tobacco mosaic virus, and for this reason smoking or use of tobacco products should be avoided around tomatoes. Various forms of mildew and blight are also common tomato afflictions, which is why tomato cultivars are usually marked with letters like VFN, which refers to disease resistance to *verticillium* wilt, *fusarium* fungus, and nematodes.

Some common tomato pests are cutworms, tomato hornworms, aphids, cabbage loopers, whiteflies, tomato fruitworms, flea beetles, slugs, and Colorado potato beetles.

Pollination

In the wild, original state, tomatoes required cross-pollination; they were much more self-incompatible than domestic cultivars. As a floral device to reduce selfing, the pistils of wild tomatoes extended farther out of the flower than today's cultivars. The stamens were, and remain, entirely within the closed corolla.

As tomatoes were moved from their native areas, their traditional pollinators, (probably a species of halictid bee) did not move with them. The trait of self-fertility (or self-pollenizing) became an advantage and domestic cultivars of tomato have been selected to maximize this trait.

This is not the same as self-pollination, despite the common claim that tomatoes do so. That tomatoes pollinate themselves poorly without outside aid is clearly shown in greenhouse situations where pollination must be aided by artificial wind, vibration of the plants (one brand of vibrator is a wand called an "electric bee" that is used manually), or more often today, by cultured bumblebees.

The anther of a tomato flower is shaped like a hollow tube, with the pollen produced within the structure rather than on the surface, as with most species. The pollen moves through pores in the anther, but very little pollen is shed without some kind of outside motion.

The best source of outside motion is a sonicating bee such as a bumblebee or the original wild halictid pollinator. In an outside setting, wind or biological agents provide sufficient motion to produce commercially viable crops.

Hydroponic and Greenhouse Cultivation

Tomatoes are often grown in greenhouses in cooler climates, and indeed there are cultivars such as the British 'Moneymaker' and a number of cultivars grown in Siberia that are specifically bred for indoor growing. In more temperate climates, it is not uncommon to start seeds for future transplant in greenhouses during the late winter as well. With the transplanting of tomatoes, there is a process of hardening that the plant must go through before being able to be placed outside inorder to have greater survival.

Hydroponic tomatoes are also available, and the technique is often used in hostile growing environments as well as high-density plantings.

Picking and Ripening

Tomatoes are often picked unripe (and thus green) and ripened in storage with ethylene. Ethylene is a hydrocarbon gas produced by many fruits that act as the cue to begin the ripening process. Tomatoes ripened in this way tend to keep longer but have poorer flavour and a mealier, starchier texture than tomatoes ripened on the plant. They may be recognized by their colour, which is more pink or orange than the other ripe tomatoes' deep red.

In 1994 Calgene introduced a genetically modified tomato called the 'FlavrSavr' which could be vine ripened without compromising shelf life. However, the product was not commercially successful and was only sold until 1997.

Recently, stores have begun selling "tomatoes on the vine", which are determinate varieties that are ripened or harvested with the fruits still connected to a piece of vine. These tend to have more flavour than artificially ripened tomatoes (at a price premium), but still may not be the equal of local garden produce.

Slow-ripening cultivars of tomato have been developed by crossing a non-ripening cultivar with ordinary tomato cultivars. Cultivars were selected whose fruits have a long shelf life and at least reasonable flavour. These have been nicknamed "Thrushworthy Bumbletots", as this is the name of the machine system used to clean the tomatoes before processing to be canned or shipped. The name is an example of onomatopoeia.

Modern Uses of Tomatoes

Tomatoes are now eaten freely throughout the world, although their seeds cannot be digested and pass straight through human intestines. Today, their consumption is believed to benefit the heart. Lycopene, one of nature's most powerful antioxidants, is present in tomatoes and has been found to be beneficial in preventing prostate cancer, among other things. The greatest benefits are accrued through eating cooked tomatoes rather than raw.

Botanically a fruit, the tomato is nutritionally categorized as a vegetable. Since "vegetable" is not a botanical term, there is no contradiction in a plant part being a fruit botanically while still being considered a vegetable.

Tomatoes are used extensively in Mediterranean and Middle Eastern cuisines, especially Italian ones. The tomato has an acidic property that is used to bring out other flavors. This same acidity makes tomatoes especially easy to preserve in home canning as tomato sauce or paste. The first to commercially can tomatoes was Harrison Woodhull Crosby in Jamesburg, New Jersey. Tomato juice is often canned and sold as a beverage. Unripe green tomatoes can also be used to make salsa, be breaded and fried, or pickled. The town of Bunol, Spain, annually celebrates La Tomatina, a festival centred on an enormous tomato fight. Tomatoes are also a popular "non-lethal" throwing weapon in mass protests; and there is a common tradition of throwing rotten tomatoes at bad performers on a stage, although this tradition is more symbolic today.

Known for its tomato growth and production, the Mexican state of Sinaloa takes the tomato as its symbol.

Culinary uses of tomatoes include:

- Tomato paste
- Tomato puree
- Tomato pie
- Gazpacho (Andalusian cuisine)
- Ketchup
- Pa amb tomaquet (Catalan cuisine)
- Pizza
- Tomato sauce (common in Italian cuisine)

Storage

Most tomatoes today are picked before fully ripe. They are bred to continue ripening, but the enzyme that ripens tomatoes stops working when it reaches temperatures below 12.5 °C. Once an unripe tomato drops below that temperature, it will not continue to ripen. Once fully ripe, tomatoes can be stored in the refrigerator but are best kept and eaten at room temperature. Tomatoes stored in the refrigerator will lose flavour, but will still be edible; thus the "Never Refrigerate" stickers sometimes placed on tomatoes in supermarkets.

Myths of the Tomato

There are many legends about the tomato. For example, it has been claimed that tomatoes were not widely eaten in the

U.S. until the late 1800s. It has sometimes been claimed that tomatoes were considered aphrodisiacs and so were shunned by the Puritans. Other claims centre on the supposed fear that tomatoes were poisonous, based on the fact that they belong to the Solanales Order, or "Nightshade" family, which contains many toxic plants. Many legends also maintain that the tomato was introduced into the U.S. from South America by one particular person; Thomas Jefferson is sometimes mentioned.

Tomatoes' status as an aphrodisiac may be due to a mistranslation. Legend has it a Frenchman on his travels ate a meal with tomatoes in it and was fascinated with the new taste. He went back to the chef who was Italian and asked him what this new ingredient was. The chef said "Pomme de' Moors" (Apple of the Moors), but the Frenchman misunderstood and thought he said "Pomme d'Amore" (Apple of Love). The modern Italian word for tomato is "pomodoro."

The most famous legend of this sort was introduced by Joseph S. Sickler in the mid-1900s, and became the subject of a CBS broadcast of *You Are There* in 1949. The story goes that the lingering doubts about the safety of the tomato in the United States were largely put to rest in 1820, when Colonel Robert Gibbon Johnson announced that at noon on September 26, he would eat a basket of tomatoes in front of the Salem, New Jersey, courthouse. Reportedly, a crowd of more than 2,000 persons gathered in front of the courthouse to watch the poor man die after eating the poisonous fruits, and were shocked when he lived. In his book Smith notes that there is little, if any, historical evidence for any of these legends, and that they continue to be repeated largely because they are entertaining stories.

It is also said that the tomato became popular in France during the French Revolution, because the revolutionaries' iconic colour was red; and at one point it was suggested that they should eat red food as a show of loyalty. Since European royalty was still leery of the nightshade-related tomato, it apparently was the perfect choice. This may also be why the first reported use of the tomato in the U.S. was in New Orleans, Louisiana, in 1812, because of the French influence in that region. There is also a story which claims that an agent for Britain attempted to kill General George Washington by feeding him a dish laced with tomatoes during the American Revolution.

"Tomato" also has been used a slang word for an attractive woman. This use was most common from the 1920s through the 1940s.

Controversies

Botanical Classification

In 1753 the tomato was placed in the genus *Solanum* by Linnaeus as *Solanum lycopersicum* L. (derivation, 'lyco', wolf, plus 'persicum', peach, *i.e.*, "wolf-peach"). However, in 1768 Philip Miller placed it in its own genus, and he named it *Lycopersicon esculentum*. This name came into wide use but was in breach of the plant naming rules. Technically, the combination *Lycopersicon lycopersicum* (L.) H.Karst. would be more correct, but this name (published in 1881) has hardly ever been used. Therefore, it was decided to conserve the well-known *Lycopersicon esculentum*, making this the correct name for the tomato when it is placed in the genus *Lycopersicon*.

However, genetic evidence (*e.g.*, Peralta & Spooner 2001) has now shown that Linnaeus was correct in the placement of the tomato in the genus *Solanum*, making the Linnaean name correct; if *Lycopersicon* is excluded from *Solanum*, *Solanum* is left as a paraphyletic taxon. Despite this, it is likely that the exact taxonomic placement of the tomato will be controversial for some time to come, with both names found in the literature.

A genome project for the tomato was begun in 2004, with a draft version of the full genome expected to be published by 2008. The genomes of its organelles (mitochondria and chloroplast) are also expected to be published as part of the project.

Fruit or Vegetable

Botanically speaking, a tomato is the ovary, together with its seeds, of a flowering plant: a fruit or, more precisely, a berry. However, from a culinary perspective, the tomato is not as sweet as those foodstuffs usually called fruits and it is typically served as part of a main course of a meal, as are other vegetables, rather than at dessert. As noted above, the term "vegetable" has no botanical meaning and is purely a culinary term.

This argument has led to actual legal implications in the United States. In 1887, U.S. tariff laws that imposed a duty on vegetables but not on fruits caused the tomato's status to become a matter of legal importance. The U.S. Supreme Court settled this controversy in 1893, declaring that the tomato is a vegetable, using the popular definition which classifies vegetable by use, that they are generally served with dinner and not dessert. The case is known as *Nix v. Hedden* (149 U.S. 304). Strictly speaking, the holding of the case applies only to the interpretation of the Tariff Act of March 3, 1883, and not much else. The court does not purport to reclassify tomato for botanical or for any other purpose other than paying a tax under a tariff act. However, the USDA also considers the tomato a vegetable.

The tomato has been designated the state vegetable of New Jersey. Arkansas takes both sides by declaring the "South Arkansas Vine Ripe Pink Tomato" to be both the state fruit and the state vegetable in the same law, citing both its botanical and culinary classifications. In 2006, the Ohio House of Representatives passed a law that would have declared the tomato to be the official state fruit, but the bill died when the Ohio Senate failed to act on it.

But due to the scientific definition of a fruit and a vegetable, the tomato still remains a fruit when not dealing with tariffs. Nor is it the only culinary vegetable that is a botanical fruit: eggplants, cucumbers, and squashes of all kinds (including zucchini and pumpkins) share the same ambiguity.

The grocers' definition is that a tomato is a vegetable based on the fact that fruits are sweet and vegetables are not.

Safety

On October 30, 2006 the U.S. Centre for Disease Control and Prevention (CDC) announced that tomatoes might be the source of a salmonella outbreak causing 172 illnesses in 18 states. The affected states include: Arkansas, Connecticut,

Georgia, Indiana, Kentucky, Maine, Massachusetts, Michigan, Minnesota, North Carolina, New Hampshire, Ohio, Pennsylvania, Rhode Island, Tennessee, Virginia, Vermont and Wisconsin. Tomatoes have been linked to 7 salmonella outbreaks since 1990 (from the Food Safety Network).

Tomato Records

The heaviest tomato ever was one of 3.51 kg (7 lb 12 oz), of the cultivar 'Delicious', grown by Gordon Graham of Edmond, Oklahoma, in 1986. The largest tomato plant grown was of the cultivar 'Sungold' and reached 19.8 m (65 ft.) length, grown by Nutriculture Ltd (UK) of Mawdesley, Lancashire, UK, in 2000.

Rapeseed

Rapeseed (*Brassica napus*), also known as **Rape**, **Oilseed Rape**, **Rapa**, **Rapaseed** and (one particular variety) **Canola**, is a bright yellow flowering member of the family Brassicaceae (mustard or cabbage family). The name is derived through Old English from a term for turnip, *rapum*. Some botanists include the closely related *Brassica campestris* within *B. napus*.

Cultivation and Uses

Rapeseed is very widely cultivated throughout the world for the production of animal feed, vegetable oil for human consumption, and bio-diesel; leading producers include the European Union, Canada, the United States, Australia, China and India. In India, it is grown on 13% of cropped land. According to the United States Department of Agriculture, rapeseed was the third leading source of vegetable oil in the world in 2000, after soybean and oil palm, and also the world's second leading source of protein meal, although only one-fifth of the production of the leading soybean meal. World production is growing rapidly, with FAO reporting that 36 million tonnes of rapeseed was produced in the 2003-4 season, and 46 million tonnes in 2004-5. In Europe, rapeseed is primarily cultivated for animal feed (due to its very high lipid and medium protein content), and is a leading option for Europeans to avoid importation of GMO products.

Natural rapeseed oil contains erucic acid, which is mildly toxic to humans in large doses but is used as a food additive in smaller doses. Canola, originally a syncopated form of the abbreviation "Can.O., L-A." (Canadian Oilseed, Low-Acid) that was used by the Manitoba government to label the seed during its experimental stages, is now a tradename for low erucic acid rapeseed that is sometimes mis-applied to other varieties.

The rapeseed is the valuable, harvested component of the crop. The crop is also grown as a winter-cover crop. It provides good coverage of the soil in winter, and limits nitrogen run-off. The plant is ploughed back in the soil or used as bedding. On some ecological or organic operations, livestock such as sheep or cattle are allowed to graze on the plants.

Processing of rapeseed for oil production provides rapeseed animal meal as a by-product. The by-product is a high-protein animal feed, competitive with soya. The feed is mostly employed for cattle feeding, but also for pigs and chickens (though less valuable for these). The meal has a very low content of the glucosinolates responsible for metabolism disruption in cattle and pigs. Rapeseed "oil cake" is also used as a fertilizer in China, and may be used for ornamentals, such as Bonsai, as well.

Rapeseed leaves and stems are also edible, similar to those of the related bok choy or kale. Some varieties of rapeseed (called, you cai, lit. "oil vegetable" in Chinese; *yu choy* in Cantonese; and *nanohana* in Japanese) are sold as greens, primarily in Asian groceries.

Rapeseed is a heavy nectar producer, and honeybees produce a light coloured, but peppery honey from it. It must be extracted immediately after processing is finished, as it will quickly granulate in the honeycomb and will be impossible to extract. The honey is usually blended with milder honeys, if used for table use, or sold as bakery grade. Rapeseed growers contract with beekeepers for the pollination of the crop.

Nutritional Value

Canola oil (or rapeseed oil) contains both omega-6 and omega-3 fatty acids in a ratio of 2:1 and is only second to flax oil in omega-3 fatty acid. It is one of the most heart-healthy oils and has been reported to reduce cholesterol levels, lower serum tryglyceride levels, and keep platelets from sticking together. Some UK farmers (such as Farrington Oils) have started to produce cold-pressed rapeseed oil as a versatile cooking oil and dressing, similar in use to olive oil.

Bio-diesel

Rapeseed oil is used in the manufacture of bio-diesel for powering motor vehicles. Bio-diesel may be used in pure form in newer engines without engine damage, and is frequently combined with standard diesel in ratios varying from 2% to 20% bio-diesel. Formerly, due to the costs of growing, crushing, and refining rapeseed bio-diesel, rapeseed derived bio-diesel cost more to produce than standard diesel fuel. Prices of rapeseed oil are at very high levels presently (start November 05) due to increased demand on rapeseed oil for this purpose. Rapeseed oil is the preferred oil stock for bio-diesel production in most of Europe, partly because rapeseed produces more oil per unit of land area as compared to other oil sources, such as soy beans.

Rapeseed and Health

Rapeseed has been linked with adverse effects in asthma and hay fever sufferers. Some suggest that oilseed pollen is the cause of increased breathing difficulties. This is unlikely however, as rapeseed is an entomophilous crop, with pollen transfer primarily by insects. Others suggest that it is the inhalation of oilseed rape dust that causes this, and that allergies to the pollen are relatively rare. There may also be another effect at work; since rapeseed in flower has a distinctive and pungent smell, hay fever sufferers may wrongly jump to the conclusion that it is the rapeseed that is to blame simply because they can smell it.

Controversy

The Monsanto Company has genetically engineered new cultivars of rapeseed that are resistant to the effects of its herbicide Roundup. They have been vigorously prosecuting farmers found to have the *Roundup Ready* gene in Canola in their fields without paying a license fee. These farmers have claimed the *Roundup Ready* gene was blown into their fields and crossed with unaltered Canola.

Other farmers claim that after spraying Roundup in non-Canola fields to kill weeds before planting, *Roundup Ready* volunteers are left behind, causing extra expense to rid their fields of the weeds.

In a closely followed legal battle, the Supreme Court of Canada found in favour of Monsanto's patent infringement claim for illegal growing of *Roundup Ready* in its 2004 ruling on Monsanto Canada Inc. v. Schmeiser. The case garnered international controversy as a court-sanctioned legitimation for the global patent protection of genetically modified crops.

There was also major concern that the extensive use of herbicide led to significant loss of biodiversity as wildflowers are killed, leaving other wildlife dependent on the wildflowers unable to survive. These concerns are offset by the reduced need to till soil during cultivation when herbicide resistant crops are used, which promotes the conservation of topsoil.

Temperate Vegetables

Cole Crops: It is one of the most important group of vegetable crops which is widely grown and popular in almost all the regions of the country. The word 'Cole' is abbreviated from `Caulis' which means stem. Probably, the stem being quite prominent, the term 'Cole' was used to refer the group of these plants originating from a single wild form, namely *Brassica Oleracea* var. *Sylvestris*. This group constitutes mainly cabbage, cauliflower, knolkhol, Brussels sprouts and sprouting broccoli. The Mediterranean Sea coast is considered. The centre of origin of these crops from where they spread first in Europe and then to almost all the countries of the world ranging from temperate to tropical.

Among the various Cole crops, cabbage and cauliflower are most widely grown on commercial scale in India. While crops like knolkhol, Brussels sprouts and sprouting broccoli are grown on very small scale and their area is scattered. They are preferred for growing in Kitchen garden.

Roots—Tubers: The tropical origin root crops are cultivated in different climatic zones of the sub-tropics because of the fact that it needs shorter growth periods like potato and sweet potato and capable for tuberization in longer day lengths and lower temperatures. Daylength is also important but it varies between 6 to 12 hours. The short day conditions are highly favourable for economic yield of vegetables and tuber crops. The different varieties are suitable for different day length and temperature.

However, adaptability in different micro-climatological regions needs suitable varieties for different environmental conditions and capacity for growing in different ecosystems. Thus physiological characters play an important role in adaptation. The day length plays an important role in tuberization of root crops. To secure good yield for Potato a day length of 12-13 hours is needed and night temperature 20^{0}C increase tuberization. With short day length and temperatures sweet potato can be cultivated in sub-tropics round the year.

Soil water is necessary for germination and establishment of crops and this applies to root crops where vegetative propagation in used. The Yams also has a high correlation between rainfall, development of vine and final yield. The sweet potato needs 50-100 cm rainfall per year and yield is reduced even when drought occurs for a month or so but water logging is also not desirable. Tapioca however withstands low rainfall of 30-50 cm. Potato needs 60-75cm but it must be evenly distributed every week during growth and development period for better yield.

It has been shown by research workers that the photosynthesis was increased with increase of CO_2 concentration and photosynthesis was closely related to the potassium content of the leaves and maximum photosynthesis occurs when 4% or more potassium are available. Similarly with high leaf area photosynthesis is more.

The yield of root crops grown during summer season in sub-tropical condition is generally lower due to long days and high temperature and also the high night temperature and as a result there is slow tuber initiation and growth while high humidity increases pests and diseases. But during winter all the conditions are favourable for higher yield.

It is well known that nitrogen increases dry matter by increasing leaf area and thus there is a relationship between nitrogen content and leaf area. But in root crops high nitrogen increases top growth and leaf area, while net assimilation decreases. Root yield depends mostly on potassium. High potassium does not change leaf area index but increases net assimilation rate and thus high dry mater.

In short, the favourable conditions for higher yield of tropical root crops in the sub-tropical conditions are favourable nutrient levels, light condition, photosynthetic activity, adaptability and soil conditions for endorsement of tuberous roots.

Bulbs and Tubers

In horticulture, the word bulb includes underground modified stems which are used for propagation *e.g.* bulb, corm, tuber and rhizome. Plants with tuberous roots are also grouped as bulbous plants. A large number of these plants producing attractive flowers are grown in the hills and are commercially important plants in floriculture. Many types, flower well both in plains and hills but the season of growth and flowering may vary.

A bulbous plant has normally three phases during a year—growth, flowering and dormancy. Vegetative growth may precede flowering as in Gladiolus or flowering starts before the leaf emerges, *e.g.* Amaryllis, Haemanthus. After growth and flowering, the plants in most cases, enter into rest period and the duration of dormancy varies with the type of plants and environmental conditions like temperature and humidity.

Cultivation

Bulbs prefer loam or sandy loam soil. In stiff clay, rooting is delayed and too much moisture often causes rotting. If the soil is not perfectly well-drained, the bulbs may be planted on a bed of sand. Watering is not required after planting as the fleshy underground stems contain sufficient food materials to develop initial growth of root and shoot. Before root formation, watering prove injurious to the bulbs and helps in rotting.

Bulbs, corms, tubers and rhizomes are used for vegetative propagation. Various methods and structures used for propagation are as follows.

- Division of clump—Rhizomatous plants the Canna, Alpinia can be divided into small clumps.
- Offsets produced laterally are separated for multiplication.
- Cormlets or bulblets produce new plants.
- Scales of Lilium are separated and used for propagation.
- Pieces of bulb disc with leaves or rhizome also develop new plan
- Bulbil arising from the axils of leaves can produce new plants.

Bulbous plants are grown in beds, shrubbery, herbaceous border, grassy land and in pots and bowl. Bulbs, corms, rhizome or tuber are usually planted when they have shown signs of sprouting after dormancy.

The planting materials are normally placed deep in the soil and the soil around it is gently pressed. The depth of planting varies with the type or plants and size of bulbs. For most of the bulbous plants grown in tropical garden, the planting depth is between 3-10 cm.

Most of these plants sunny situation, while Eucharis, Zephyranthes thrive better in semishade; few types, *e.g.* Pancratium, Haemanthus, Zephyranthes grow and flower well in sun and semishade. Bulbs, corms and tubers should not come in direct contact with fresh organic manure, rhizomes are not usually affected. Time of planting depends on the season of flowering, environmental condition of the region and condition of the planting material. In the plains, Gladiolus is planted from September to November, whereas in the hills, the planting season is normally from April-June. Amaryllis, Hippeastrum, Haemanthus are planted in January-February to obtain flowers in March-April. Caladium and Canna are put in the ground in May.

After the above ground portion dries out the bulbs are lifted out carefully. The root and shoot are cut and the bulbs are cleaned before storage. They should be stored in dark, dry and airy place until the time of next planting. Storage condition is very important for obtaining healthy bulbs. If the humidity is very high, fungus develops on the bulbs and causes rotting. In a very dry atmosphere the bulbs shrived and may lose viability.

It is not necessary to lift all types of bulbous plants every year. Amaryllis, tuberose, Caladium, Pancratium, Zephyranthes may remain in the ground for three years before they are lifted. But water should be withheld when the bulbs are lying in dormant condition in ground or pot. Gladiolus, on the other hand, has to be lifted out every year after drying of leaves.

4

Some Common Bulbous Plants

1. *Amaryllis (Amaryllidaceae):* Amaryllis belladona is the only species of the genus, a native of South Africa. It is a plant with strap-shaped, glossy, green leaves and producing few large funnel-shaped flowers on a stout stalk, from March-May. Leaves appear at the end of the flowering season and die out before the winter sets in. In warm humid region the plants retain leaves throughout the year and bear flower during the season. Offsets, developed from the original bulb are separated for multiplication. Bulbs are planted 5-8 cm deep in December-January and the plants grow well in rich and porous soil. Amaryllis is not disturbed for several years at one place and the plants continue to produce new bulbs and flowers. It is very suitable for planting in border, shrubbery and in pot. Although rose or rosy pink colour is common, white to purplish colour are also found.
2. *Caladium (Araceae):* The genus comprises of about 16 herbaceous perennial species with tuberous rhizome. They are a native of tropical America and are widely grown in warm humid climate for the beautiful and attractive foliage produced during the rainy season. Some of the important species are Caladium bicolor, C.humboldtii, C.picturatum, and C.schomburgkii. Numerous varieties have been raised particularly form C.bicolor.

   Leaves are usually peltate-segittate, stalks variegated, blades are with very many shades of colour, *e.g.* variegated green, blue green, dark green, light green, spotted with white, red, transparent white, etc. leaf veins may be red, silvery or green. Leaf margins are coloured with purple, white, yellow or red. In some species, shape of the blade is lanceolate-sagittate. Flowers are unisexual.

The pot compost for planting Caladium should be prepared by mixing loamy soil with leafmould and well-decomposed cow dug manure. It prefers moist but porous soil. Application of liquid manure at frequent intervals enhances growth and improves colour of foliage. Caladium is propagated through division of the tuberous rhizome and rarely by seeds. It is planted in summer and leaves show a fine display of colour in the rainy season.

The leaves start fading in autumn and water is gradually withheld until they have withered. In dormant condition, the rhizome may be allowed to remain in the soil or dug out and stored in a cool and dry place.

3. *Cooperia (Amaryllidaceae):* The genus is named after Joseph Cooper, an English gardener, comprises about 6 species native of North America, and differing from Zephyranthes by the long perianthtube and erect anthers. They are tender bulbous plants with the habit of Zephyranthes but blooms only at night. Important species are C.drummondii and C.pedunculata. Flowers are fragrant, white, and sometimes tined with red or pink.

   Flowers are solitary, the perianth subtended by a bract like spathe. Leaves are long, narrow flat and twisted and appear along with the flower. It is cultivated in a semishady location and in sandy loam soil. Addition of well-rotted compost, sand and charcoal dust in beds or pots is beneficial. Cooperia is propagated by bulbs. It is planted during spring in rock garden, border or in pots and starts blooming during May-June.

4. *Cooperanthes (Amaryllidaceae):* This genus is a product of intergeneric hybrid between species of Cooperia and Zephyranthes, first raised in 1900 by Percy Lancaster at the Agri-Horicultural Society, Alipore, Calcutta. Hybrid between Cooperia oberwettii x Zephyranthes robusta is

known as Alipore Beauty, is probably the best known Cooperanthes. Flowers are light lilac, rose or white.

It grows well in well-drained sandy loam soil, rich in organic matter and in semishade. Cooperanthes are propagated by bulbs. They are planted in the spring and the plants start flowering by the end of summer. They are grown both in bed or in pot and are good as cut flowers.

1. *Crinum (Amaryllidaceae):* The genus Crinum, comprises more than 100 species of large and showy flowering bulbous plants. They are closely allied to Amaryllis and distinguished by the longer perianth tube. The species cross freely and many fine hybrids of Crinum, between Crinum and Amaryllis and Crinum and Hymenocallis are known. Some of he important cultivated species and hybrids are C. longifolium, C. mooriei, C. powellii, C. variabile, C asiaticum, C. augustum, C. careyanum, C. yuccaeflorum, C. giganteum, C. zeylanicum. Flowers are usually white or in shades of red and purple.

   The stems arise from the tunicated bulbs with a more or less elongated neck. Leaves are large, about 150 cm long and 12-15 cm wide, evergreen or deciduous, depending on the species. Flowers are regular, often highly scented, tube narrow, with six segments and usually funnel-shaped. In some species flowers are 30 cm long and 15 cm wide.

   Most of the species prefer shade or semishady location for planting. Crinums usually have large bulbs, sometimes as much as 60 to 80 cm long with numerous fleshy roots. If planted in beds, the soil should be dug to a depth of two to three feet, mixed up well with sufficient quantity of rotten cow dung and compost. It can also be grown in large pots containing soil rich in organic matter.

   Bulbs should be of good size and planting is done to a depth below the ground level, twice the size of the bulbs. The plant will grow and flower for years, if watered in the summer months and top dressed with fresh loamy soil around new vegetative growth. Bulbs are planted in April to get flower during the rainy season.

2. *Eucharis Grandiflora (E. amazonica) (Amaryllidaceae)::* Eucharis is an important bulbous plant, popularly known as 'Amazon Lily'. It prefers semi shade and flowers better in plains than on the hills. The bulbs are globular in shape, leaves large lanceolate. The flowers are about 7 cm across, white, sweet, scented, 5-7 blooms appear on a stalk. Eucharis grows better in pots than in the ground and the compost should be rather rich and heavy instead of sandy, but must not be often disturbed. It flowers in the summer and rains and the leaves begin to wither before the winter.
3. *Gladiolus (Iridiaceae):* Gladioli are among the most beautiful flowers, blossoming from October to March in plains and during June to September in the Hills. This genus comprises more than 150 species of perennial herbs with base of stem swollen into a corm. Most of them have their origin in South Africa, although a few originated from Europe. The species in cultivation and of considerable importance include: G. cardinalis, G. childsii, G. colvillei, G. gandavensis, G.lemoinei, G. nanceianus, G. nanus, G. primulinus, G. psittacinus, G. purpureoauratus, G. saundersii. The numerous crosses have evolved large number of varieties of varying characters and it is almost impossible to keep up with and enumerate the different varieties of gladioli. Wide range of attractive colour shades may be grouped into white, yellow, cream shaded pink, orange, pale pink on white, pink on cream apricot, salmon, cherry red, orange-scarlet, crimson, purple, mauve and violet, etc.

   Gladioli prefer sunny situation with light sandy soil, with pH between 6 to 7. If the soil is heavy, addition of river sand and charcoal improves the soil condition. They grow well both in pot and in beds and the magnificent spikes brighten the garden and room as cut flower. Two parts sandy loam soil mixed with one part of each well-rotted cow dung manure and leafmould and a handful of bone meal is recommended for pot culture. If planted in beds the ground should be dug deep and well-decayed compost at the rate of 4 kg per meter mixed thoroughly and left for sun drying at least for a fortnight. At planting time fresh manure should not be used, but a dressing of bonemeal, superphosphate and wood ash may be given with advantage. In the plants, planting is usually done during September-November and in the hills from April-June, but the flowering season may be extended by early and late planting.

Gladioli are propagated by seeds, corms and cormlets. New varieties are raised from seeds. Seeds germinate freely and the seedlings grown carefully, will flower in the second year. Large flower spikes, however, develop after 3-4 years when the corms attain a good size. Gladioli corms are planted 8 to 10 cm deep into the soil, at a distance of 20-30 cm between the rows and the plants. Application of liquid manure, once at the vegetative stage and again after the formation of flower buds has been found very effective.

The flowering spikes appear in 60-90 days after planting, the flowers continue to open in succession from below upwards and the open flowers remain fresh for a number of days. After flowering, the leaves begin to turn yellow and wither. The plants are then lifted with the corms and cormlets and kept in a dry shady place for a week for drying. The corms are thoroughly cleaned, the cormlets separated and stored on a layer of sand in a dry, cool, airy and shady place until the next planting season. The corms should be examined regularly and those showing sign of rotting or fungus growth should be removed. They are also stored in cold storage but in a dry atmosphere.

4. *Gloriosa (Liliaceae):* As the name implies, the genus Gloriose means 'full of glory', popularly known as climbing or creeping lily. They are tall creeping plants, support themselves by means of tendril which arises from leaves. This genus comprises about six species of rhizomatous plants native of Africa and tropical Asia. Important species, commonly grown in the gardens in India are G. superba and G. rothschildiana. Leaves are oblong,

lanceolate. Flowers are showy on long pedicels in leaf axils, perianth of 6 distinct long segments; stamens six with versatile anthers.

Well-drained soil in a sunny location is ideal for planting Gloriosa and the attractive flowers make a fine display of colour when trained on bamboo frame work or low trellis. Gloriosa is vegetatively propagated from rhizome, which may be cut in pieces and planted to a depth of 3-4 cm in April-May.

The plant flowers in July-August. After flowering is over the plant begins to wither and rhizome becomes dormant. If left undisturbed. Gloriosa continues to produce flowers for several years in the same place. Application of liquid manure once at the active phase of vegetative growth and another just before flowering is recommended.

5. *Haemanthus (Amaryllidaceae):* Haemanthus is one of the popular and attractive bulbous plants, commonly known as Football Lily or Bloody Lily. The genus comprises nearly 60 species, native of South and tropical Africa. The flowers are red, crimson, scarlet, pink and some species are white or pale green in colour. Some of the important species are H. multiflorus, H. magnifica, H. lindenii, H. albo-maculaius, H. coccineus, H. tigrinus, and H. candidus.

   Bulbs are usually large with thick skin. Stems green, short, thick and fleshy. Leaves are usually large and luxuriant, turn yellow and dry in the winter months, the scape is sometimes curiously coloured. Inflorescence is a dense, many flowered umbel, perianth straight and erect with a short cylindrical tube. Flowers are showy and produced in ball-like heads. Fruits berry like, indehiscent. Flowers often appear before the leaves or sometimes simultaneously.

   To plant in beds, the soil is deep dug and mixed with well-rotten compost and planted 40 cm apart. It grows well in pot and small pot is preferred. Application of organic manure in the rains and liquid manure before flowering is beneficial and the plants will continue to flower for several years in the same place.

They are propagated by offsets, which should be detached from the mother plant during spring. The bulbs are planted in pots or in beds during spring season to get bloom in summer and early part of rainy season.

6. *Hedychium (Zingiberaceae):* Hedychium consists of 40 species of rhizomatous herbs native of Asia and several species are grown for their beautiful and fragrant flowers in both plains and hills. It prefers semi-shade and moist soil and produces many flowered large spikes during July-October. The leaves die before winter.

7. *Hemerocallis (Liliaceae):* Hemerocallis is popularly known as Day Lily as the flowers last for a day and the blossoms fade at night. This genus consists of more than a dozen species and they are mostly native of China and Japan. The plants are very hardy, stout rooted, glabrous, perennial herbaceous plants, admired for their showy blooms.

   The leaves are almost grass like, 2 ranked at the base of the scape. Flowers are lily like, large funnel shaped, yellow or reddish orange or brown in colour. Though the individual flower last for a day, but many flowers open successively to keep the lasting beauty for a long period.

   Hemerocallis grows in wide variety of soil. While planting, the ground should be dug to a depth of 50 to 60 cm and mixed up with well-rotten compost. They take a year or two to establish properly and should not be disturbed very frequently. Planting is done in February to March and flowering continues from May to August.

   Propagation is by division of the clumps and also by seeds. Some of the species are self-sterile and seeds can be obtained easily by crossing two species. Many new-varieties with attractive large flowers of various colours have developed by hybridisation. After 4-5 years, Hemerocallis clumps may be forked out during January-February separated and can be replanted again in the planting season.

8. *Hippeastrum (Amaryllidaceae):* Hippeastrum, commonly known as Royal Dutch Amaryllis is one of the finest

flowering bulbous plants. The flowers of Royal Dutch Amaryllis are larger, open widely, about 15-20 cm across, tube is shorter and the plant develops larger bulb than Amaryllis belladona. The spectacular flowers have various shade of bright colour and wide range of variation exists in varieties developed by hybridisation. Hippeastrum produced by Ludwig are famous throughout the world. A double flowered variety is also available.

The leaves are broader than Amaryllis and do not dry out in the plains. The bulbs are lifted in November-December, stored in dry cool place for a few weeks and replanted in January-February. Hippeastrums develop flower during March-April and application of liquid organic manure before emergence of flower stalk improves the size and quality of flowers.

9. *Hymenocallis (Amaryllidaceae):* This genus comprises about 40 species of bulbous plants native of South America (one in Africa *i.e.* H. speciosa) cultivated for beautiful fragrant flowers, consisting of narrow green white petals with large coronas and protruding stamens. Some of the important species are H. macrostephana, H. harrisiana, H. speciosa, H. caribaea, H. calathina, H. rotata, H. amancaes, H. tubiflora. Flowers are white, except H. amancaes which is bright yellow.

   Well-drained soil in sunny location is good for planting Hymenocallis. The bulbs are planted just below the surface of the soil during February to get the bloom in the month of May. The soil should be prepared by mixing 2 parts of loamy soil, one part of each well-rotten cow dung manure and leafmould. Watering should be given freely during the growing season. So long the plants remain healthy, transplanting or shifting from pot to pot is not recommended. Whenever necessary bulbs are dug out during September –October and stored in dry place for planting in February next year. Hymenocallis is propagated by bulbs.

10. *Nelumbo (Nymphaeaceae):* Nelumbo or Nelumbium, popularly known as Lotus is a flower of national

importance in India. The genus consists of two species– the one bearing yellow flower and the other produces white or pinkish flower.

The plant has rhizomatous stem, large peltate leaves standing well above the ground and it spreads rapidly in shallow water or wet soil. Like Nymphaea, Lotus prefers soil rich in organic matter and flowers profusely in summer and rains. Lotus do not grow well in lily pools and so tank is ideal for it. N. lutea bears scented yellow flowers 15-20 cm across N. nucifera is common in India. The flowers are white or pink, 20-25 cm in diameter. Lotus is propagated by rhizome or seed.

11. *Nymphaea (Nymphaeaceae):* The Water Lily is the most popular flowering plant in water garden. The genus comprises of 40 species of aquatic, rhizomatous plants widely grown in tank, lily pool, streams, lakes and in suitable earthenware containers.

The leaves are oval or round in shape, floating on the water or standing above the water surface. The flowers are very attractive, show various colours: white, yellow, pink, red, blue and shades of colours are found in the varieties developed by hybridisation.

Water Lily prefers clay soil rich in organic matter and sunny location. The rhizomes are planted in soil in the bed of the tank or lily pool or in baskets or pots filled with rich clay soil and then placed in the tank in March-April. In the gardens in cities and towns water lily is often planted in large earthenware container filled with water and a layer of soil. In several species and varieties do not develop in the winter months and the rhizomes remain dormant. Vegetative growth starts in spring and flowers appear during the summer, rains and early winter.

In some species flowers open in day time while in other blooms open at night and close during the day. The species of Nymphaea commonly grown in water garden are as flowers: N. caerulea produces light blue flowers, 7-10 cm across, day blooming. N. lotus, the night blooming water lily, bearing white flower 12-15 cm across. N. odorata, a white flowering and fragrant water lily. The flowers are 8-15 cm across. N. pubescens, has white flowers, 8-12 cm across. N. stellata, a species commonly found in the warm humid region of India. It produces pale blue flowers about 10-16 cm in diameter. Nymphaea is propagated from rhizomes, bulbils and seeds. Most of the species and varieties thrive and flower well at water depth between 50 to 150 cm.

12. *Polyanthes Tuberosa (Amaryllidaceae) Tuberose:* Tuberose (Polianthes tuberosa), a native of Mexico, is widely grown in the plains of India and blooms profusely during the summer and rains, flaunting its fragrance outdoors and indoors. Most artistic garland, floral ornaments, bouquets and buttonholes are made from these flowers. The long spike of flowers is excellent for table decoration. The flowers remain fresh for days together and bathe the atmosphere with their sweet pleasant fragrance.

    The tuber is bulb-shaped and the plant is commonly classed among the 'Bulbs'. The leaves are 70-80 cm long, narrow, linear and radical, bright green in colour. The flowering stalk which emerges from the centre of the cluster of leaves is about 80-120 cm long bearing successively smaller long pointed clasping leaves, uppermost ones are much reduced and bract-like. The flower buds are tubular. Flowers are 5-6 cm long, borne in pairs in an open spike pure wavy white, highly fragrant, tube 2.5 to 3.0 cm long, slightly bent near the base, expanding widely where it meets the oblong obtuse segments.

    There are three types of tuberose in cultivation - 'single' with one row of corolla segments; 'semi-double' bearing flowers with two to three rows of segments and 'double' having more than three rows of corolla segments. Though there is no popular named variety in tuberose. 'The Pearl' is known to be a variety in the double flowered type. A variety with variegated leaf bearing single flowers is also grown. Single flowered type is more widely cultivated than the other types.

The bulbs remain dormant during the winter months in places where the temperature is low and if early planting is desired, the dormancy can be successfully broken by dipping the bulbs in 4% thiourea solution for one hour. Normally tuberose begins to flower in 90 –95 days after planting. It flowers during the summer and rains (April-September) in the plains of eastern part of the country and from May to July on the hills, while in milder climate tuberose flowers well throughout the year.

Bulbs having diameter 2.0 and 2.5 cm show satisfactory growth and flowering. The average life of a flower spike is about 10-15 days in situ, while that of an individual flower varies from 4-6 days. Vase life of a spike varies from 7-10 days, depending on the environment and change of water. The spikes remain fresh for a longer period, if kept in 4% sugar solution.

Tuberose can be successfully grown in pots, beds, borders, shrubberies and rockeries. Propagation is by means of seeds and bulbs. Vegetative propagation is commonly practiced and desirable too, because such plants produce better flowers within a short period after planting.

The land should be thoroughly cultivated until the soil comes to good tilth. A good amount of rotted cow dung or farm yard manure should be incorporated with the soil at least 10-15 days before planting. The bulbs are planted 4-5 cm deep in beds in March-April in the plains and in May-June on the hills. The planting distance between the bulbs is 10-15 cm, while the rows are spaced 25 cm apart. For pot cultivation, 1-2 bulbs are planted in a 20 cm pot.

A fertilizer mixture containing 6 gms of urea, 16 gms each of superphosphate (single) and muriate of potash per square meter has been found to show satisfactory growth and flowering. The above mixture should be applied in two equal doses – the first dose before planting and the second one, 4 weeks after sprouting of the bulb. Commercial growers dig out bulbs 3 years after planting. They are stored in a cool, dry and shady place and planted again in the following spring.

13. *Zephyranthes (Amaryllidaceae)::* Zephyranthes has a fanciful meaning *i.e.* the 'West-wind Flower' commonly known as Zephyr Flower or Fairy Lily. The Genus has about 50 species of bulbous flowering plants, native of warmer region of America. It is related to Habranthus, Pyrolirion and Hippeastrum. Some of the popular species are Z. candida, Z. citrina, Z. grandiflora, Z. rosea, Z. macrosiphon, Z. tubispatha and Z. Verecunda. The flowers are with many colour ranges *i.e.*, white, yellow and various shades of pink.

Zephyranthes is a hardy bulbous herb. Leaves are filiform or linear or may be strap shaped. Flowers solitary, peduncle elongated slender, hollow; perianth funnel-shaped, erect or slightly inclined, stamen 6, stigma 3-fid.

It is commonly propagated by separating the bulbs, sometimes through seeds. Bulbs are planted in spring, 30 to 40 mm deep and 2 to 15 cm apart. They flower during summer and rains and make a very colourful display especially in rains. Top dressing of organic manure once a year in a rainy season will promote flowering and the bulbs may be left undisturbed for many years.

Potato

Group: Root Vegetables

Photo: A pile of potatoes

Latin Name: Solanum Tuberosum

Popular Varieties: Estima, Maris Piper, Charlotte, King Edward, Desiree

Also Known As: Taters, Spuds

Storage: Potatoes should be stored between 5 and 12°C.

Comes From: Worldwide; potatoes do best in temperate countries, moderate rainfall is required to swell tubers.

Seasonality: All year round

You could say Potatoes are one of the five most important foods in the world, and you would be right. The Food and Agriculture commission confirmed this when in 2005, it recorded worldwide production as being 322 million (metric) tons – that's a whole lot of potatoes! They are the fourth most important source of food energy in the world after rice, wheat and maize, and are the worlds' most important tuber crop.

For that is what potatoes are; a tuber. A tuber is different to a root; it is an underground storage organ, in which the potato stores energy in the form of carbohydrate; primarily starch. The balance of sugars to starches is what varies tastes and uses of potatoes between varieties.

There are several main "classes" of potatoes; these were originally white round, russetted, and red, but has recently been extended to include speciality and gourmet, to include any varieties that don't fit in the original three. However, all potatoes are more commonly split depending upon their optimum use: baking (they retain their shape when baked), roasting (they have good flavour when roasted), boiling (they hold some shape when boiled), salad (usually firm and waxy when boiled) or mashing (they mash into a smooth pulp without fibres or grains). Potatoes are not limited to just white and red varieties, you can get yellow, purple, blue and even black skinned potatoes!

Most people know not to eat green potatoes, because they are poisonous. However what most people don't realise is all potatoes are technically poisonous, but don't run off scared yet!

Potatoes contain compounds called glycoalkaloids, the two primary ones being solanine (which you may have heard of) and chaconine. They are both metabolic poisons, present in the greatest concentrations underneath the skin; their concentration increases with exposure to light. The green in potatoes is not these glycoalkaloids, instead it's chlorophyll, which also develops in the presence of light, which is why the green is a good indication of the presence of glycoalkaloids. It is not a definitive guide however, since they can occur independently of each other. Cooking at high temperatures (over 170°C or 340°F) partly destroys them, which is why boiling does not!

Why are all potatoes poisonous then? This is because these glycoalkaloids are present in all potatoes, but not in toxic quantities; 200mg of solanine is classed as a dangerous dose, which could be received by eating one medium sized green potato, or more than 10 good potatoes (about 3 pounds or 1.5kg) at one time. Potatoes are now grown in over 125 countries worldwide, and as such have gathered quite a collection of names; the original name "potato" came from the Spanish "patata", in turn coming from the American Indian word "Batata". In Ireland they are known as spuds (a term derived from a tool used to harvest them) or taters, and in Scotland as taters or murphies. In America, they are sometimes known as "Irish Potatoes" to differentiate them from Sweet Potatoes which whilst appearing similar, are actually completely unrelated and are in fact a storage root rather than a tuber.

Potato storage is tricky, but easy to get the hand of. They should not be refrigerated: this causes the starches in the potatoes to be converted to sugars, giving them an unnaturally sweet taste. This sugar will caramelize during cooking producing brown potatoes with poor taste. However at room temperature they will only last for a week or two before losing too much moisture and shrivelling; they should ideally be stored between 5-12°C, in a dark place to prevent sprouting.

History

Approximately 7,000 years of potato history has been tracked down, starting in Peru and the Andes in 5,000BC and spreading from there. Pre-cursors of the Inca civilisation originally cultivated it and it originally grew in Native South American tribes. It's first appearance in Europe was in 1537, when the Spanish conquistador Castellanos discovered the tuber when raiding a village in South America, and brought it back to Spain. It is thought they were introduced to England by Sir Francis Drake around the time of the Spanish Armada.

The potatoes spread across Europe was initially slow; people distrusted them as a food. Since the potato was not mentioned in the bible (probably due to its origins in South America rather than Eastern Europe), some people refused to eat it, with Russian peasants christening them "Devil's Apples", and many other people associating potatoes with common diseases; the French believed them to cause leprosy. It was not until the late 18th century that they finally were accepted. France largely accepted potatoes following the famine following the Seven Years' War; a French scientist Antoine Parmentier wrote of their usefulness as a famine food whilst eating them during his time as a prisoner of war in Prussia; after the war he established soup kitchens in Paris largely based around potatoes. Parmentier also created French fries, serving them at a dinner solely based around potatoes to honour Benjamin Franklin, who was not particularly impressed; it was a later president, Thomas Jefferson, who introduced French Fries to America..

Potatoes were however introduced to America earlier than this, one record dates this to around 1719 in New Hampshire, by Scottish-Irish settlers, although earlier documents record the Governor of Bermuda, Captain Nathaniel Butler, sending a cargo of potatoes to Francis Wyatt, the governor of Virginia in 1621..

The one nation who bought into the virtues of potatoes early paid dearly for it. Potatoes arrived in Ireland not long after England (it's thought that wreckage from ships of the Spanish Armada inadvertently brought potatoes to Irelands' shores), and before long it became a primary crop. This was due mainly to their high productivity, if the Irish could survive on a crop that took up less land, there would be a greater area free for wheat production for export. By 1650 potatoes were the staple food of Ireland and a large part of its national diet. They also began to replace wheat as the major crop throughout Europe, for feeding both people and animals.

However mother nature is not always on our side. In 1845 and 1846, a late outbreak of potato blight, a fungal disease that devastates the growing top of potatoes and turns the tubers to mush, ripped through Europe, wiping out the crop in many countries. This had devastating consequences in Ireland; it is estimated 1 million people died during the Irish Potato Famine, and millions more emigrated, largely to Northern America. Ireland continued to export crops during the famine, but the high price of the exported food meant the poor Irish peasants could not afford to buy it for themselves.

Uses

Potatoes are a very high energy food source; they have a high carbohydrate content in the form of starch and sugars, and also contain protein and vitamins (particularly vitamin C). However, the method of preparation determines just how much of this goodness is retained.

The obvious use for potatoes is boiling them; either peeled or unpeeled for new varieties with softer skins, whole or cut into pieces, with or without seasoning. Cooking is largely required for any preparation, breaking down the starch and making the potatoes edible. Once boiled, they can be served whole or mixed with butter and mashed or creamed. Steaming is an alternative to boiling, which retains more of the nutrients.

Baked potatoes are also popular; the potato is usually (but not always) baked in its skin, and served with varied accompaniments, ranging from plain cheese, to maiyonnaise, tuna, chicken, chilli con carne or others, depending on what is provided where it is prepared. Potatoes are also good roasted in the oven, a good serving companion to roast chicken, or beef.

Potatoes can be sliced and fried; these are usually termed chips (in England) or French fries elsewhere, although these can also be oven baked (which will retain less fat and is more healthy). Potatoes can be cut into thin strips and fried (hash browns or rosti/potato pancakes), or formed into dumplings.

Another popular use is potato chips (or crisps in England), where thin strips of potato are fried or roasted until dry and then flavoured. As you can see, the number of food uses for potatoes is simply astronomical.

Pea

A pea is the small, edible round green bean which grows in a pod on the leguminous vine *Pisum sativum*, or in some cases to the immature pods. This legume is cooked as a vegetable in many cultures. Several other seeds of the family Fabaceae, most of them round, are also called peas; this article deals with the species *Pisum sativum* and its cultivars. The pea plant is an annual plant, with a lifecycle of a year. The average pea weighs between 0.1 and 0.36 grams.

Peas are a cool-season vegetable crop. The seeds may be planted as soon as the soil temperature reaches 10 °C, with the plants growing best at temperatures of 13 °C to 18 °C. They do not thrive in the summer heat of warmer temperate and lowland tropical climates, but do grow well in cooler high altitude tropical areas. Many cultivars reach maturity about 60 days after planting. Peas grow best in slightly acid, well-drained soils.

Peas have both low-growing and vining cultivars. The vining cultivars grow thin tendrils from leaves that coil around any available support, and can climb to be 1-2 m high. A traditional approach to supporting climbing peas is to thrust branches pruned from trees or other woody plants upright into the soil, providing a lattice for the peas to climb. Branches used in this fashion are called *pea brush*. Metal fences, twine, or netting supported by a frame, are used for the same purpose. In dense plantings, peas give each other some measure of mutual support.

"The pea ranks among the oldest grain legumes of the Old World" write Daniel Zohary and Maria Hopf (Zohary & Hopf 2000, p.101). They document the earliest find-sites as being located in southern Syria and southeastern Turkey, and elaborate that peas "seem to be associated with the spread of Neolithic agriculture into Europe ... closely associated with the wheat and barley production" (Zohary & Hopf 2000, p. 106).

Diseases

Ways of Eating Peas

In early times peas were grown mostly for their dry seeds. Along with broad beans and lentils these formed an important part of the diet of most people in Europe during the Middle Ages (Bianchini 1975 p 40). By the 1600s and 1700s it became popular to eat peas "green", that is, while they are immature and right after they are picked. This was especially true in France and England, where the eating of green peas was said to be "both a fashion and a madness" (OSU 2006). New cultivars of peas were developed by the English during this time which became known as "garden peas" and "English peas." The popularity of green peas spread to North America. Thomas Jefferson grew more than 30 cultivars of peas on his estate (Kafka 2005 p 297). With the invention of canning and freezing of foods, green peas became available year-round, not just in spring as before. Fresh peas are often eaten boiled and flavoured with butter and/or spearmint as a side dish vegetable. Salt is also commonly added to peas when served. Fresh peas are also used in pot pies, salads and casseroles. Pod peas are used in stir fried dishes, particularly those in American Chinese cuisine. Pea pods do not keep well once picked, and if not used quickly are best preserved by drying, canning or freezing within a few hours of harvest.

Dried peas are often made into a soup or simply eaten on their own. In Japan and other Southeast Asian countries including Thailand, Taiwan and Malaysia, the peas are roasted and salted, and eaten as snacks. In the UK, marrowfat peas are used to make pease pudding (or "pease porridge"), a traditional dish. In North America a similarly traditional dish is split pea soup.

In Chinese cuisine, pea sprouts (*dou miao*) are commonly used in stir-fries and its price is relatively high due to its agreeable taste.

In the United Kingdom, dried, rehydrated and mashed marrowfat peas, known by the public as mushy peas, are popular, originally in the north of England but now ubiquitously, and especially as an accompaniment to fish and chips or meat pies, particular in fish and chip shops. Sodium bicarbonate is sometimes added to soften the peas. In 2005, a poll of 2,000 people revealed the pea to be Britain's 7th favourite culinary vegetable. Processed peas are mature peas which have been dried, soaked and then heat treated (processed) to prevent spoilage — in the same manner as pasteurising.

Cooked peas are sometimes sold dried and coated with wasabi as a spicy snack.

Some forms of etiquette require that peas be only eaten with a fork and not pushed onto the fork with a knife.

Peas in Science

Etymology

According to etymologists, the term was taken from the Latin *pisum* and adopted into English as the mass noun *pease*, as in pease pudding. However, by analogy with other plurals ending in -*s*, speakers began construing *pease* as a plural and constructing the singular form by dropping the "s", giving the term "pea". This process is known as back-formation.

The name *marrowfat pea* for mature dried peas is recorded by the OED as early as 1733. The fact that an export cultivar popular in Japan is called *Maro* has led some people to assume mistakenly that the English name *marrowfat* is derived from Japanese.

Plantation Crops

Rubber

Malaysia, Indonesia, Thailand and Africa are the main rubber producing countries. Natural rubber for commercial production is available from Manihot glaziovii (cera rubber), Ficus elastica (India rubber), Castiolla elastica (Panama rubber), Parthenium argenatum (Guayul), Taraxacum koksaghyz and Hevea brasiliensis (Para rubber) and among them, Hevea brasiliensis is the most important commercial source of natural rubber. It is native of Brazil and was introduced in Asia in 1876. After proper chemical treatment, rubber wood provides enough strength and durability of any semi-hard wood available in India and can be used for the manufacture of useful articles like door and window components, furniture, wall panelling, interior decoration, tool handles etc.

Rubber tree belongs to the natural order Euphorbiaceae. This tree is sturdy, tall and quick growing. It has a well developed tap root and laterals. The leaves are trifoliate, with long petioles. Flowers are unisexual, small and fragrant. Staminate flowers are small and numerous. Pollination is by insects. Latex vessels are present in all parts of the tree except in the wood.

Climate and Soil: Rubber exacts in its climatic requirements. The regions lying within 10^0 latitude on either side of the Equator is highly suitable for rubber cultivation. It requires a temperature ranging from 20^0 to 30^0C with a well distributed rainfall of 200-250cm over the year. It comes up in plains and also in slopes of mountainous regions ranging from

300-800m above sea level. This specific climate is available only in Kanyakumari district, Tamil Nadu and Kerala, which constitute the traditional area. It thrives well in deep well drained acidic soils of red lateritic loams or clayey loams with a pH varying from 4.5 to 6.0.

Varieties: Rubber Research Institute of Malaya, Rubber Research Institute of India, kottayam and other institutes have developed clonal varieties. These clones are broadly classified into three categories *viz.*, primary, secondary and tertiary, based on the method adopted for the development of their mother trees. When mother trees are selected from existing seedling populations of unknown parentage and are multiplied vegetatively to give rise to the clones, they are called primary clones. When the mother trees clones and are then multiplied vegetatively, they are known as secondary clones.

Rubber Board of India recommends some of the following clones for cultivation in South India.

	Name	Parents	Characters
	Primary clones		
1.	TJIR - 1	-	Indonesian clone, yield 930 kg/ha per year, Susceptible to Phytophtora, Oidium and pink disease.
2.	G.T.1	-	Indonesian clone, yield 1360 kg/ha year, Tolerant to Phytophthora, pink disease and brown bast.
3.	G.I.1	-	Malaysian clone, yield 1130 kg/ha year, Susceptible to brown bast, possess drought tolerance.
4.	P.B. 86	-	Malayasian clone, yield 1130 kg per ha/year.
	Secondary clones		
5.	PRIM-600	TJIR-1 x P.B. 86	Developed by Rubber Research Institute of Malaya (RRIM). Yield 1317kg/ha/year, Susceptible to Phytophthora and pink disease.
6.	RRIM. 628	TJIR.1x RRIM.527	Yield 1051 kg/ha/year, Susceptible to brown bast, poor yielded during summer.
	Tertiary clones		
7.	RRIM - 703	RRIM.600xRRIM.500	Yield 1725 kg/ha/year, Susceptible to brown bast and wind damage.

Other important clones recommended for South Indian conditions are PR.107, PB.5/51, RRIM.118, RRII-203 and RRII –208.

Propagation

(i) *Seeds:* Propagation through seed is practised to raise seedlings for rootstock purpose or to raise polyclonal seedling progenies. Seeds normally ripen during July-September in South India. As the viability is very short (8 weeks), they are to be sown immediately raised beds of river sand of 1m width and of convenient length are formed and the seeds are sown in a single layer toughing one another and pressed firmly with the surface of the seed just visible above. Nursery may be protected from direct sun by providing a temporary shade. Regular watering is attended to maintain the moisture in the beds. Seeds start germinating within 6 to 10 days. Such raise seedling stumps or at 60x90cm or 60x120cm to raise bud wood nursery or stumped budding. Otherwise, sprouted seeds can be directly planted in the field.

(ii) *Budding:* The scions of a particular clone is maintained in the bud wood nursery by planting the budded stumps or by budding the clone on the seedlings in situ at nursery. Budded stump often refers to the budded plant whose scion shoot is cut very close to the budding zone leaving few dormant buds in the scion shoot. On the other hand, if the root stock is cut as a stump and budding is done, usually green budding at four to five months stage, then it is known as stumped budding.

When the budwood nursery plants are one year old, about 1m of usable budwood can be obtained. The budwood is cut when atleast 1m of brown bark has developed. The immature green portion should be removed to a point about 1m below the terminal bud, leaving the leaf stalks in position. The budwood may be cut off about 15cm at the base, leaving a few dormant buds to develop into bud shoots for the subsequent season. Two such sprouting shoots may be allowed for next year, from one metre shoot, 15 to 20 buds may be obtained.

Modified forket method is followed and is done during April-May, when the weather is not dry or wet. Two types of budding techniques are practiced. Brown budding is done by using buds taken from bud wood of one year growth on to a stock plant of ten months old. Green budding on the other hand involves young green budwood and stock. Bud wood of 6-8 weeks old is used on stock seedlings of 2 to 6 months old.

Recently, polybag plants are raised as such plants reach tapping stage quickly. Black polythene bags of 60x30cm with 400 guage are filled with topsoil alone along with 25g of rock phosphate. Green budded stumps are planted in these polybags and the scions are allowed to develop 2 to 3 whorl of leaves.

Planting: For new plantings, jungle clearing with felling of trees has to be done first. Pits are usually dug to the size of the 1x1x1m^3 and are filled up with soil and compost. Planting may be done in a rectangular or square or quincunx system during June-July. The common spacings adopted for budded plants are

In hilly areas - 6.7 x 3.4 m

In flat areas-square - 4.9m x 4.9m

Triangular - 4.9m x 4.9m

Interculture

Cover cropping: Growing cover crops is important in rubber plantations to prevent soil erosion, conserve soil moisture, keep down the soil temperature and add mulch and organic matter to the soil. Some of the cover crops commonly used in South India are Pueraria phaseoloides, Calopogonium mucunoides, Centrosema pubescens and Mimosa invisa var. inermis.

It is often desirable to establish a mixed cover to simultaneously get the benefits of these cover crops. A mixture of Calopogonium, Pueraria and Centrosema seeds in the ratio of 5:1:4 is used for sowing. In this mixed cover, Calopogonium grows much rapidly and covers the ground quickly during the first year itself. Then Pueraria and Centrosema start dominating. Dense and vigorous growth of Pueraria suppresses weeds. But it starts fading out when the canopy closes. Subsequently, Centrosema continues to grow under shaded conditions while the former two cover crops fail to thrive. Thus, complete benefits of cover crops starting from the first year of planting could be obtained if mixed cover crop is established in rubber planations.

Weeding: The weeds can be eradicated either by labour or by employing weedicides. When weedicides are employed, care should be taken that the cover crops are not affected. 2,4-D formulations (Fernoxone @2kg in 450 litres of water) may be sprayed early in the season to eradicate the weeds.

Manures and Fertilisers

Three stages of growth namely nursery, immature and mature can be recognised in the life of a rubber tree. The manuring differs according to the stages of growth. The Rubber Research Institute in India recommends the following manurial schedule.

Seedling nursery: Application of 25kg of compost and 2.5kg of rock phosphate once in three years per 100m^2 of the nursery bed is practised. Application of 25 kg of 10:10:4:1.5 NPKMg mixture per 100m^2 of the nursery bed 6 to 8 weeks after planting and application of 12.5kg of the mixture per 100m^2 6-8 weeks after the first application but before mulching are followed.

Budwood nursery: Powdered rock phosphate 1.5kg per 100m^2 of the nursery bed is applied as a basal dressing at the time of preparing the nursery bed. Besides 250g of 10:10:4:1.5 NPKMg mixture per plant in two split doses of 125g each, the first dose is applied two to three months after planting the budded stumps or cutting back, if budding is carried out in situ and the second dose eight to nine months after planting.

Immature trees at pretapping stage:

Year of Planting	No. of applications	Time of application	Dose of NPKMg mixture (10:10:4:1.5) per plant (g)
First	1	Sept-Oct	225
Second	2	April-May Sept-Oct	450 450
Third	2	April-May Sept-Oct	450 550
Fourth year onwards till tapping	2	April-May Sept-Oct.	550 550

Rubber Trees under Tapping: Application of NPK 12:6:6 grade mixture at the rate of 400kg per hectare per year in two split doses is recommended.

In plantation where the trees show deficiency symptoms of magnesium (interveinal yellowing of leaves) addition of 10kg commercial magnesium sulphate to every 100kg of the mixture is recommended during September-December.

Apping

The rubber trees attain tappable stage in about seven years provided they possess the required girth of the trees. Seedling must attain a girth of 55cm at a height of 50cm from the ground. In the case of budded trees the girth should be 50 cm at a height of 125–150cm from the bud union. Tapping is the periodical removing of thin slices of bark to extract rubber latex. Tapping is done by skilled men. While tapping the depth should be 1mm close to cambium without any damage to it, otherwise callus formation will take place causing swellings.

Tapping has to be done on a slope of 30^0 to the horizontal zone in the case of budded trees and 25^0 in the case of seedlings. Tapping is done early in the morning, as late tapping will cause reduction in the flow of latex. In the early morning the turgor pressure in the latex vessels is high and rapid flow of latex occurs.

Tapping System: The following tapping systems are generally followed in India.

Tapping system	Intensity	Remarks
S_2d_1 – Half spiral, daily tapping.	200%	Followed by small growers but it favours brown bast incidence and causes early deterioration of trees.
S_2d_3 – Half spiral, tapping at every three day for 6 months and rest for 3 months.	67%	Recommended for clonal seedlings.
S_2d_2 – Half spiral, tapping alternate days for 6 months and rest for 3 months.	100%	Recommended for budded plants

In South India, rubber trees shed their leaves during December-January and immediately again they put forth new leaves and flowering. During this period the trees are given rest since the yield of rubber will be poor if tapped. The yield of rubber steeply increases year by year and the peak is reached 14-18 years after planting. Then it slowly declines. After 40 years it may not be economical to maintain the trees. The latex yield will vary with the clone, age of the trees, fertility of the soil, climatic conditions and skill of the tapper. In the case of old trees, tapping may be done intensively adopting a system of two half spirals, one at normal basal level and the other at a higher level on the opposite side and away from first one atleast 120 – 180 cm.

Intensive tapping prior to feeling of the old trees is called slaughter tapping. It is often done at higher levels, sometimes even one branches with the help of ladders and not on the usual renewed bark levels. As the objective of slaughter tapping is to extract as much latex as possible from the available bark, no consideration is given to the technique, intensity or standard of tapping. Tapping is not done on rainy days but by fixing a polythene rain guard to the trunk of the tree above the tapping panel, tapping can be carried out during rainy season also. About 35-40 additional tapping per annum can be obtained by rain guarding the trees under the alternative daily system. It is recommended in areas where the annual yield is 700kg/ha or more and where normally more than 25 tapping days are lost by rain.

A number of chemicals have been shown to influence the flow of latex after tapping, among which, ethrel (2-chloroethy1 phosphonic acid) has been found to stimulate and increase the yield of rubber latex two-fold. Ethrel has to be diluted with coconut oil to have 10% active ingredient and is applied thrice during the year *i.e.*, March, August, September and November.

Yield: In South India, the annual yield of rubber is about 375kg per hectare per annum from the seedlings, whereas budded plantations yield 900 to 1000kg of rubber/ha.

Processing of Rubber: The latex that flows out from the rubber trees on tapping is channelled into a container, generally coconut shell cups, attached to them. Latex collected in coconut shell cups in transferred to clean buckets, two to three hours after tapping. The latex which gets dried up on the tapping panel (tree lace) and the collection cups (shell scrap) also form a part of the crop and are collected by the tapper in baskets just prior to tapping. The latex spilt including overflows on the ground (earth scrap), when gets dried up, is also collected once in a month. Normally 10-20 % of the total crop constitute the tree lace, shell scrap and earth scrap. Rubber can be processed and marketed as:

Preserved Latex Concentrates: The latex is collected in the storage tank, from there it is brought to a centrifuge machine, rotating at 1440 rpm. Due to the centrifugal action, liquid portion comes out. The upper layer, the concentrated latex, is collected and brought to bulking tank and mixed with chemical and packed in drums. 60% rubber is present in it. Skim latex is taken to another tank and sulphuric acid is added and coagulated and milled to get skim crepe. It is of poor quality while the concentrated latex fetches very higher price.

Dry Ribbed Sheet Rubber: Anti-coagulants (solutions of aminonia, formalin or sodium sulphite) are added to the cups to prevent the coagulation of latex before it reaches the factory. The latex so collected is bulked and then strained to remove the impurities. It is then diluted to a standard consistency of 12-13% rubber. Special hydrometers like metrolac, latex meter are employed to measure the percentage of rubber. After dilution, the latex is strained through a 60 mesh screen for the second time. Then it is poured into the special coagulating tanks or aluminium pans which is divided into many compartments by thin aluminium sheets and acetic acid or formic acid to used for coagulation. Slow coagulation produces a soft rubber, which is easy to work on the rollers.

The acid is to be added quickly and mixed thoroughly with the surface of rubber sheets. After coagulation, rubber sheets are repeatedly washed several times with changes of water and passed through hand or power operated rollers. In the roller excess water and dissolved impurities are pressed and sequeezed out. The surface of the rollers may be either smooth or grooved or zig zag or straight or diamond pattern, its impression is normally left on the surface of the sheets when they come out of the press. These sheets are hung in shade for two to three hours for dripping in a dust free place. They are then taken to smoke houses for thorough drying. Smoking of rubber sheets is done to dry the sheets properly and to avoid formation of blisters. In the smokehouse, the sheets are smoked at a low temperature of 48-50^{0}C with fairly high humidity during the first day subsequently during 2nd to 4th day the temperature being 68^{0}C with low relative humidity. They are taken out, graded and packed. Such products are known as smoked sheets or dry ribbed sheet rubber. Various grades of rubber sheets are RMA IX, RMA-1, RMA-2, RMA-3, RMA-4 and RMA-5. High grade rubber sheets are clear, free from blisters, translucent and of a golden brown colour and fetch a better price.

Dry Crepe Rubber: When coagulum from latex or any form of field coagulum after necessary preliminary treatments is passed through a set of creping machines to get crinkly, lace-like rubber called 'crepe rubber' after drying. Various grades of crepe rubbers are EPC super 1 X, EPC1X, EPC2X, and EPC3X.

Plant Protection: The important diseases and pests of rubber and their contorl measures are

Disease/Pest	Symptoms	Control measures
Abnormal leaf fall (Phytophthora palmivora)	Infected leaves fall in large numbers prematurely.	Spray Bordeaux mixture (1%) as prophylactic measure, prior to the onset of South West monsoon.
Powdery mildew (Oidium haveae)	Ashy coating noticed on tender leaves	Dusting 3 to 6 rounds at 10-15 days interval using 11-14 kg of 325 mesh fine sulphur dust per round per hectare.
Scale insects (Saissetia nigra)	Severly affected portion dry up and die.	Spray malathion at 0.05% concentration.
Mealy bug (Perrisiana virgata)	Severly affected portion dry up and die.	Spray malathion at 0.05% concentration.

The oldest known beverage, tea is native of China in South East Asia. It was known to the Chinese as early as 2737BC, but attained the status of a popular drink in England in 1664 AD. India is the largest producer, consumer and exporter in tea industry. Tea belongs to the genus *Camellia* and family Camelliaceae. The original species, which produces tea, were *C.assamica* (Assam jats), *C.sinensis* (China jats) and their natural hybrid, *C.assamica* subspecies lasiocalyx (Indo China or cambod type). Being a highly cross pollinated crop, the present day seedling populations are mixture of both the above two species, however, from their major share of characters–Assam or China type can be distinguished by the following characters.

Assam	China
It is a tree	It is a shrub
Few robust branches	Branches abundant and whippy
Large, glossy leaves	Small leathery leaves.
Light to medium green	Dark green colour
High yield and medium quality	Low yield but good quality
Susceptible to drought and frost	Hardy and resistant
Sparse flowering	Profuse flowering

Morphologically, tea is an evergreen shrub or tree, leaves are simple, alternate, serrate, flowere bisexual, with superior ovary, fruit is a capsule.

Varieties: Clonal selection from seedling population was taken up by UPASI, Tea Scientific Department, Cinchona and also by other Tea Research Institutes. UPASI has so far released 27 clones. Certain outstanding clones released by other Institutes are also used in South India.

Clone	Characters
URASI-2 (Jayaram)	An average yielding clone, suitable for all elevations, tolerant to drought and wind.
URASI-3 (Sundaram)	Very high yielding and quality clone.
URASI-6 (Brooklands)	Fares well at mid and higher elevations.
URASI-8 (Golconda)	High yielding, suitable for all elevations.
URASI-9 (Athrey)	High yielding, fairly tolerant to drought, Can withstand slightly high pH.
URASI-10 (Pandian)	Hardy clone, resistant to drought and wind; suitable for high elevation.
URASI-14 (Singara)	Quality clone, suitable for higher elevations.
URASI-17	A high yielding clone.
TRI-2024 (Sri Lanka)	High yielding clone.
URASI-2025 (Sri Lanka)	Average yielding;hardy clone.

Tea is exacting in its climatic requirements. The temperature may vary from 16 to 320C and annual rainfall should be 125 to 150 cm, which is well distributed over 8-9 months in a year. The atmospheric humidity should be always around 80% during most of the time. Very dry atmosphere is not congenial for tea. It is grown in plains in North Eastern States but in South India, it is grown in hill ranges from 600 to 2200 m above M.S.L.

Tea is a calcifuge crop requiring comparatively low amounts of calcium but high quantities of potassium and silicon. They can be grown in lateritic, alluvial and peaty soils. Optimum pH range is 4.5 to 5.0 and soil depth should be 1.0 to 1.5m.

Propagation: Tea can be propagated by seed and by cuttings. Seeds collected from the fruits of seed baries are soaked in water and only heavy seeds, which sink, are alone used for sowing in beds. Germination occurs in 20 to 30 days. At that stage they are carefully lifted and transplanted in polythene sleeves. They will be ready for planting in 9 months.

Vegetative propagation: The site for the nursery can be selected in a flat land or gentle slope, near to a perennial water source and easily accessible by road. It should have a good drainage and should be protected from wind, frost and wild animals etc. approximately, 0.15 ha nursery area is required to produce 1.25 lakhs cuttings. Nursery area is to be provided with overhead shade by erecting concrete or stone pillars at a spacing of 3x3m and spread with 6mm^2 mesh double strand coirmat which provides about 67% shade.

The cuttings for rooting are collected from mother bushes, which are well maintained near the nursery area. Such mother bushes are pruned well in advance to induce juvenile shoots. These juvenile shoots are collected in the morning hours and 3cm long cutting each with a healthy mother leaf and an active axillary bud is prepared. Cuttings from top tender and bottom brown wood should be avoided. These cuttings are planted in polythene bags (30cmx10cmx150 gauge), filled with growing medium (Jungle soil: river sand 3:1) in the bottom and rooting medium (Red/subsoil:sand 1:1) in the top 8-10cm. The soil used for rooting media should have an optimum pH range of 4.8 to 5.0, if high, *i.e.*, 5.1 to 5.5, or 5.6 to 6.0, it must be drenched with 1 or 2% aluminum sulphate solution respectively @ 1 litre per cubic foot of soil. This treatment should follow with drenching of twice the volume of plain water to wash excess aluminum sulphate. The cuttings are carefully planted at the centre of the bags in such a way that the petiole should not touch the soil and then they are watered. These bags are then covered with polythene sheets over the G.I. wire arhes and the sides are tugged well to preserve moisture content. Callusing starts in 4-6 weeks and rooting occurs in 10 to 12 weeks. When 80% of the cuttings have rooted, the tents are opened in stages and the overhead shade is gradually reduced to harden the plants.

Planting: The land is cleared of the roots of the fallen trees and drains are taken at suitable intervals depending upon the slope to conserve the soil. In the olden days, up and down system of planting at 1.2x1.2m are followed. Presently, contour planting either in a single hedge or double hedge system is followed.

S. No.	Type	Spacing	Population/ha.
1.	Up and down	1.2 x 1.2m	6,800
2.	Contour planting single hedge.	1.2 x 0.75m	10,800
3.	Contour planting double hedge.	1.35 x 0.75 x 0.75m	13,200

The last method has many advantages over the first two *viz.*, early and high yield, better soil conservation, less weed growth in the hedge and efficient cultural practices. Planting season normally coincides with June/July and September/ October for Southwest monsoon and Northeast monsoon areas. Pits of 30 x 30 x4 5cm size are dug and plants of 12-15 months old are planted by removing the polythene sleeves. Immediate after planting, plants are staked to prevent wind damage.

After Care: Immediately after planting, the soil surface around the plants should be mulched, usually cutgrasses of gautemala are employed for this purpose. About 25 tonnes of grass is required to mulch one hectare. Care must be taken to keep the mulch materials away from the collar region last they may cause collar diseases. If there is a dry weather, mud tubes or etah tubes may be buried 15cm deep near the plant in a slanting position and one litre of water per plant may be poured or injected at weekly intervals. This subsoil irrigation helps to minimise the causality besides encourages developing deeper roots.

Shade Management: Tea requires filtered shade and if it is exposed to direct sun, its growth is affected. Shade is hence essential and beneficial to tea as

- It regulates the temperature.
- It minimises the effects of drought and radiation injury.
- It increases the soil fertility
- It helps in recycling of nutrients.
- It helps in getting even distribution of crop.
- It serves as windbreak.
- It reduces the incidences of pests.
- It generates additional income by way of timber and fuel.

The desirable characters of a good shade tree like

- It must be an evergreen tree, easy to propagate having quick growing and deep rooted characters.

- It provides filtered shade and withstands frequent lopping.
- It tolerates wind and frost.
- It does not have allopathic effect.
- It has commercial timber value also.

Weed Control

Weeds will be a problem in young and pruned fields. Manual weeding is never recommended in tea lest more soil erosion and damage to surface roots and collar regions. Therefore, the following chemical weed control is alone recommended in tea.

Type of weeds	Herbicides	Dosage
Dicots	Paraquat (gramoxone)	1.12 lit. /ha.
Dicots	Sodium salt of 2,4-D (Fernoxone)	1.4 kg. /ha.
Grasses	2,2-Dichloro propionic acid (Dalapon)	5.6 kg. /ha.
	Glyphosate	2.3 lit. /ha.

Training and Pruning

In the young tea, when it has established well, centering *i.e.* removing the growing point leaving 8 to 10 mature leaves from the bottom, is done to induce secondaries. When the secondaries reach more than 60 cm, they are tipped at 50-55 cm height by removing 3 to 4 leaves and bud to induce tertiaries. Therefore, plucking at mother leaf stage is continued for better frame development. It takes nearly 18 to 20 months from planting to reach regular plucking field stage.

Pruning is done in tea :

- to maintain to convenient height for plucking
- to induce more vegetative growth
- to remove dead and de funct wood and
- to remove the knots and interlaced branches.

Pruning is normally done 4 to 6 years interval depending upon the altitude of the garden, nature of the materials etc. the bushes marked for pruning should have adequate starch reserves in roots otherwise the sprouting following pruning should have adequate starch reserves in roots otherwise the sprouting following pruning will be less. This can be normally tested by the common Iodine test and if the starch reserve is less, bushes are allowed to rest for 2 to 3 months. The different types of pruning are as follows:

Sr. N.	Type of pruning	Pruning height (cm)	Season	Remarks
1.	Rejuvenation pruning	20 – China Jat 30 – Assam Jat	April - May	Done is old bushes affected with cankar and wood rot to invigorate the new healthy branches. Not done regularly.
2.	Hard pruning	30 – 45	Apr. – May	First formative pruning done to a young tea.
3.	Medium pruning	45 – 60	Aug. – Sept.	Normal pruning whereever frames are healthy.
4.	Light pruning	60 – 65	Aug. – Sept.	Normal pruning whereever frames are healthy.
5.	Skiffing	65	Aug. – Sept.	Mainly to postpone pruning and to encourage better frame development.

Immediately after the rejuvenation or hard pruning, the cut ends are smeared with a paste made of copper oxychloride and linseed oil (1:1). The prunings, consisting of only small twings and leaves are buried in trenches of 30cm width and 45cm depth taken across the slope in alternate rows. The pruned bushes are given washing with 10% lime solution using No. IV nozzle of power sprayers in order to kill the epiphytic growth of moss and lichen so as to induce early and even bud break. Lime washing also minimises sunscorch to the bush frame. The buds from the pruned shoots grow in a steady succession without any cessation of growth. These are known as a periodic shoots or primary shoots. These primary shoots should be induced to produce flush shoots, otherwise known as periodic shoots by regular tipping operation. Tipping is the removal of terminal portion of the shoot and it varies with jats and pruning height as given below. Tipping height refers to the number of leaves that must be left above the pruned cut while tipping in material refers to that portion of the terminal shoot, which must be tipped off.

Pruning height (cm)		Tipping height (cm)		Tipping in material	
China Hybrid	Assam/ Assam Hybrid	China Hybrid	Assam/ Assam Hybrid	China Hybrid	Assam/ Assam Hybrid
35-45	35-55	5	4	3 leaves and a bud	4 leaves and a bud
45-55	55-60	4	3	4 leaves and a bud	4 leaves and a bud
55-75	60-75	2	2	4 leaves and a bud	4 leaves and bud
-	> 75	-	1	-	4 leaves and a bud

Manures and Fertilisers

Tea responds to manuring and it has been estimated that to produce 100kg of made tea, tea plant utilises on an average 10.2, 3.2 and 5.4kg of Nitrogen, Phosphorus and Potash per ha. Manuring in tea starts from nursery stage itself. Once they strike roots (after 4 months) 30g of soluble mixtures (Ammonium phosphate (20:20) 35 parts, potassium sulphate and Magnesium sulphate each 15 parts and zinc sulphate and Magnesium sulphate each 15 parts and zinc sulphate – 3 parts) is dissolved in 10 litres of water and is applied with rosecan for about 900 plants. This must be repeated at 15 days intervals.

Nitrogen: The recommendation for mature tea is based mostly on soil organic matter status and anticipated yield. For a field with medium organic matter status the following rates of application is suggested for every 100kg of made tea anticipated:

Yield level (kg/ha)	Rate of Nitrogen (for 100 kg. of made tea)	No. of split applications
<3000	10 kg	4
3000	8 kg	5
3000 and above	9 kg	6

Twenty per cent of the total nitrogen is given in the form of Ammonium sulphate during March/April. Urea is recommended in May/June and receding monsoon months avoiding very wet and very dry periods and it will come to 65% of total nitrogen. Fifteen percent of the total nitrogen is applied in the form of Calcium Ammonium Nitrate during pre-winter (November-December).

Potassium: Nitrogen and potassium are always applied together. NK ratio of 1:1 is used for plucking fields while for a pruned field 2:3 NK ratio is recommended. For rejuvenation pruned field 1:2 NK ratio is suggested. The enhanced rates of potassium application in the pruned year is to encourage formation of healthy farmers. Muriate of potash is the sources of potassium used in tea fields. The NK fertilizers are applied by broadcast for mature tea and is broadcast and dibbled in along the drip circle for young tea. The interval between two successive applications should be atleast 3-4 weeks.

Phosphorus: Phosphorus is applied once in alternate years @ 90kg P_2O_5/ha for fields yielding less than 3000kg/ha for fields yielding between 3000 and 4500kg/ha, 60 to 80kg P_2O_5/ha is suggested every year. The soils being acidic, rock phosphate could be advantageously used. The fertilizer should be placed at 15-22cm depth.

Micronutrients: Among the micronutrients, zinc deficiency is often manifested in young shoots characterised by reduced leaf size, rosetting, chlorosis and formation of more banji shoots. Application of zinc sulphate @ 6 to 8kg/ha for high yielding fields every year is the general recommendation. The above quantity can be given in 4 to 5 split applications during has been found beneficial to combine other micronutirents *viz.*, Manganese sulphate @ 15.5g/10 litres and boric acid @ 5.5g/10 litres of spray volume along with zinc sulphate spray.

Liming: In the hill soils, due to the leaching of bases by rain and also due to the incessant application of acid forming fertilizers, the soil pH is often reduced which affects the physical and chemical properties of soil. Therefore, periodical application of lime is essential properties of soil. Therefore, periodical application of lime is essential to amend the soil and maintain optimum pH. Agricultural lime (Calcium carbonate) and dolomitic lime (Calcium Magnesium carbonate) are generally recommended for tea soils. The rate of application is based on soil pH, rainfall, fertilizer usage and length of the pruning cycle. Roughly lime @ 1.5mt/ha for a pH between 4.5 to 4.9, 3.0mt/ha for a pH between 4.0 to 4.4 and 4.0mt/ha for a pH of less than 4.0 is suggested.

The lime is applied by evenly broadcasting prior to pruning once in a pruning cycle. First manuring following liming can be had after 6 weeks and a minimum of 15cm rainfall should have been received during this period.

Harvesting or Plucking

Plucking consists of harvesting 2 to 3 leaves and a bud. It is the most labour intensive operation in a tea industry and also decides the yield and quality of made tea. Normally, a pluckable shoot takes 60 to 90 days for harvesting since its sprouting from the axillary buds. When the shoot is plucked upto mother leaf, it is known as light plucking and if it is plucked below mother leaf, it is called hard plucking. The plucking interval and plucking standard in relation to cropping is given below:

Cropping pattern	Months	Plucking interval
High cropping or Rush cropping (60% of total crop)	April – June and October – December	7 – 10 days
Low cropping or lean cropping (40% of total crop)	July – September and January – March	12 –15 days

It is essential to add one tier of active maintenance foliage to the bush every year. This is done by mother leaf plucking during January to March. During the rest of the period level plucking can be carried out. Consequent to plucking, bush height increases every year in the order of 10cm over tipping height in the first year, 7.5cm, 7.5cm, 5cm and 5cm over the previous year height in the second, third, fourth and fifth year respectively.

Yield

Yield of made tea per hectare depends upon many factors such as elevation, clonal or seedling jats, management practices, severity of pruning, processing techniques etc., Generally, in tea industry, a field which yields upto 2000kg of made tea/ha is considered as low yielding and 2000 to 3000kg as medium yielding and anything above 3000kg as high yielding fields.

Manufacturing of Tea

Basically, there are two types of processing *viz.*,

1. Orthodox method in which the rolling operation is done in a series of rollers. The rollers have rotary tables with battens, jacket for loading the leaf and a pressure cup,
2. CTC method (cutting, tearing and curling) which has a CTC machine, consisting of series of a pair of rollers mounted in such a way they rotate in opposite directions and the clearance between them is so adjusted to crush and tear the leaves.

Irrespective of the method, manufacturing of tea involves the following steps:

Withering: The objective of withering is to reduce the moisture content of leaves by spreading them in troughs which receive artificial air from fan fitted on one end. At the end of withering, the leaves attain a flaccid condition for which it may take 12 to 18 hours depending upon the weather condition.

Rolling: This operation is carried on by a series of machines or in a single roller, during which the cells in the leaves are broken to liberate the sap containing the polyphenol oxidase, an enzyme, which in the presence of oxygen, oxidises the polyphenols to produce theaflavins and thearubigens.

These are responsible for colouring of the tea and is a 30-40 minutes. Afterwards, the fine sifted rolled ones are sent for fermentation while the coarse ones are again sent for rolling.

Fermentation: Rolled tea materials are either spread in concrete floors or kept in aluminum trays. In the presence of high humidity and proper step decides the quality *i.e.* strength, colour and briskness of tea. Fermentation requires 1 hour or 2 hours depending upon the environmental conditions.

Drying: This step aims at stopping the fermentation process and slowly removing the moisture content without a burnt smell but preserving the inherent quality. This is achieved by passing the fermented tea in thin layers through conveyors into a drier in which the inlet temperature is maintained around 250 – 280^{0}F and outlet temperature is a round 150-200^{0}F. Proper drying takes 30-40 minutes.

Grading: Before grading, the dried tea is removed of the stalky fibres, which affect the quality, by passing through fibre separate machines. The bulk tea is passed through different sized meshes which aid in separation into different grades.

Orthodox grades	Mesh size	CTC Grades	Mesh size
Pekoe	>8 mesh sieve	Flowery Pekoe (FP)	>8 mesh
Tippy golden Orange pekɔe (TGOP)	8-12	Pekoe	8-10
Broken orange pekoe (BOP)	12-16	BOP	10-12
BOP – Fannings	16-18	Pekoe Fannings	12-16
BOP –dust	18-24	BOP – fannings	16-20
Dust – I	25-30	Pekoe dust (PD)	20-30
Dust – II	Below 30	Red Dust (RD)	30-40
		Super Reddust (SRD)	40-50
		Finel dust (PD)	50-60
		Superfine dust (SD)	Below 60

Plant Protection

Many pests and diseases are known to infect tea bushes and cause economic losses.

The important pests and diseases, their typical symptoms and control measures are:

Pests			
Sr. N	Pest	Symptoms	Control measures
1.	Tea mosquitoes (Helopeltis antonii)	Small adult bugs and hairy orange nymphs suck the sap from fresh leaves and tender shoots; leaves curl up, dry and die; active from January to September.	Collect and destroy bugs during the initial stages; spray 0.1% Malathion or 0.05% Lindane
2.	Shot-hole borer (Xyleborus fornicatus)	Grubs make a typical short-hole on the branches and inside gallaries. A serious problem in low and mid elevation areas	Badly affected branches are pruned off. Heptachlor 20 EC is sprayed @ 8.5 l in 675 lit. of water/ha on the pruned frames and prunings
3.	Red spider mite (Oligonychus coffeae)	Infests upper surface of mature leaves	Tetradifon (8 EC) 1 to 1.25lit/ha.
4.	Scarlet mite (Brevipalpus californicus)	Discolouration of leaves often leads to defoliation	Dicofol or Ethion @ 1lit. /ha.
5.	Purple mite (Calacarus carnatus)	Leaves exhibit smoky grey colour	
6.	Pink mite (Acaphylla theae)	Young leaves turn pale and get twisted.	
7.	Yellow mite (Polyphagota rosnemus latus)	Infest pluckable shoot, leaves become rough, brittle and corky in under surface.	
8.	Thrips (Scirtothrips bispinosus)	Leaf surface becomes uneven, curly and metty, exhibiting parallel lines of feeding marks on either side of the midrib	Phosalone or endosulfon 1lit/ha.
9.	Nematodes (Meloidogyne javanica, M. incognita)	Occur in tea nursery, infested roots develop galls.	Pre heat treatment of soil media upto 60-80^{0}C and application of carbofuran 3G @ 80g/m^3 of medium.

Diseases			
Sr. N	Disease	Symptoms	Control measures
1.	Blister blight (Exobasidium vexans)	Infects tender leaves and stem and develops translucent spot. Cloudy and wet weather favour infection.	Copper oxychloride 350g in 67lit of water with power sprayer for pruned field at 3-4 days interval. In the plucking Oxychloride + 210g Nickel chloride in 45lit of water/ha at 7 days interval.
2.	Black root diseases (Rosellinia arcuata)	Infested roots show black mycelium on the roots, white star shaped mycelium between bark and wood and Black lead shot like perithecia seen on collar region.	The soil may be drenched with Dithane M-45 @ 30g/10 litres.
3.	Red root disease (Poria hypolateritia)	Infected roots exhibit blood red mycelium on washing. It spreads fast but slowly kills.	• Take trenches of 1.2m deep and 45 cm width sorrounding the infected bushes and uproot and burn the bushes in situ. • Rehabilitate soil with gautemala grass. • Soil fumigation with methyl bromide carbon-di-sulphide
4.	Brown root disease (Fomes noxius)	Infected root wood turns soft and spongy, it spreads slowly but kills quickly.	

Plantation Crops

Cinchona

Cinchona is native of high lands of South America and was introduced in India (Nilgiris) in 1859. It is grown in Nilgiris and Anamalai hills of Tamil Nadu. It is also grown in Darjeeling (West Bengal). It is an evergreen tree, growing to a height of 10-12m with a sparse branching habit. It belongs to the family Rubiaceae. The important species, which are under commercial cultivation, are *Cinchona ledgeriana, C. officinalis, C. robusta and C. succiruba.*

Quinidine, an alkaloid of cinchona bark, is used for its anti-malarial, anti-pyretic and oxytonic actions. But no clinical use is currently made of these properties. In addition to their use in pharmacy, quinine and quinidine and their derivatives are utilized in insecticide compositions for the preservation of fur, feathers, wool and textiles.

Climate and Soil: Cinchona requires an average temperature of 20^{0}C with a relative humidity of 85%. Annual rainfall should be not less than 200cm but distributed atleast 8 months in a year. The best elevation is 1000 to 2000m above M.S.L. without any frost occurrence. Cinchona prefers porous, well drained, fertile soils with a thick cover of organic matter and high moisture holding capacity. The optimum pH range is 4.5 to 6.5

Propagation: Cinchona is propagated by seeds and vegetatively by cutting, stooling, layering and patch budding. Seeds are sown in raised beds during April and they take about 20-30 days for germination. The healthy seedlings are transplanted in baskets or polythene bags when they are about four months old. Clonal propagation is sometimes done through top working or patch budding.

Planting: The area selected for planting should be cleared a year in advance and planted with shade trees like Silver oak and or dedabs. Pits of 30 x 30 x 45cm are dug and filled up with topsoil and other well decomposed organic matter. Seedlings are transplanted in the main field at the spacing of 1.25 x 1.25m when they are about one year old. Transplanting is done any time when there is sufficient moisture in the field. The other method is to go for high density planting *i.e.* trees are set out at a spacing of 1.0x1.25m or 8000 plants per hectare and gradually harvested until 800 plants/ha remain after 25 years.

Manures and Fertilisers: Cinchona plants are manured with 115kg Nitrogen, 15kg Phosphorus and 115kg Potash per ha per year. Once in 3 to 4 years when the soil pH goes below 4.5 liming @ 1.0 to 1.5 t/ha is recommended

Interculture: The main cultural operations are staking the plants in the first three years to prevent wind damage. In young plantations, wind growth may be checked by slash weeding, followed by chemical weeding with paraquot @ 30ml and sodium salts of 2, 4-D @ 25g in 10 litres of water.

Harvesting: The trees are coppiced when they are 8 to 10 years old depending on the vegetative growth. Coppicing consists of pruning the tree at a height of 5cm from the ground level. The stump left on regenerates to produce a large number of shoots but only two or three of these are retained and allowed to grow further while the remaining ones are removed. The trees become ready for the second coppicing within 8 to 10 years from the time of first coppicing. After the second coppicing also, two to three shoots are left to grow further. The trees are finally uprooted in the 30th year when they start declining in vigour. Even though the major harvests are obtained at the time of first two coppicings, some yields of bark are also available from the dead and dying trees and prunings. During the first two coppicings, an yield of 4000kg of dry stem bark per hectare may be obtained and at the final stage of uprooting the tree, the yield of bark may be 6000kg per hectare. The most important alkaloid principle is Quinine which occures in the stem, twig and root bark of the tree. Normally its content range from 3 to 4% as Hydroxy Quinine sulphate.

Processing: The extraction of quinine involves beating of the bark with a mallet to loosen it for peeling by hand or knife. The peeled bark is quickly dried to prevent the loss of alkaloids. The fully dried bark is sent to the factories for solvent extraction of powdered bark with slaked lime containing more than 60%

of Calcium hydroxide and the alkaloids removed with amyl alcohol or ether.

These alkaloids are in turn extracted from the solvents in acidified water, they precipitate out when the water is made alkaline. It is then dried and powdered, and is the starting material for the manufacture of quinine base and other quinine salts. Medicinally, cinchona alkaloids (purified) and their salts are conforming to the latest pharmaceutical standards in various countries.

Plant Protection: In the nursery, damping off caused by *Pythium* is often noticed. Drenching with 0.5% copper oxy chloride is recommended at 10 days interval.

Tea mosquito bugs *(Helopeltis antonii)* often infest the leaves in the nurseries and also in the mainfield. Spraying with any systemic insecticide will check the incidence.

Plantation Crops

Cocoa

Cocoa, (Theobroma cacao) is a beverage crop introduced in India in the early 1965s. It is native of Amazon valley of South America. In India, it is cultivated mainly as a mixed crop in coconut and arecanut gardens. Kerala accounts for 79% of the total area and 71% of the total production and Karnataka shares 18% of the total area and 25% of production and the rest by Tamil Nadu. It belongs to the family Sterculiaceae.

Varieties

Commercial cocoa has two major varieties, Criollo and Forestero.

Sr. N.	Character	Criollo	Forestero
1.	Cotyledons	Plumpy and white when fresh and turn cinnomon coloured on fermentation.	Flat and purple when fresh and turn dark chocolate brown on fermentation.
2.	Pod colour	Dark red	Yellow
3.	Other pod characters	Rough surface, ridges prominent, pronounced point and thin walled	Smooth, inconspicuous ridgers, thick walled, melon shaped with rounded end.
4.	Flavour and aroma	Bland flavour	Harsh flavour, bitter taste.
5.	Duration of fermentation	3 days	6 days
6.	Adaptability in India	Poor adaptability and less yield potential and hence discourage for commercial cultivation.	Good adaptability and high yielding and hence recommended for commercial cultivation.

Other types of cocoa include: (1) Trinitario from Trinidad which is said to be a hybrid between Criollo and Forestero with highly variable pod characters (2) Amelonado, a Forestero type bean with a melon shaped pod, cultivated in West Africa and (3) Amazon, a relatively new type collected from the Amazon forests which has got vigour and high yield.

CPCRI, Kasargod recommends some selections introduced from Malaya *viz.*, I-21, II-11, II-18, II-67, III-5 and III-101 for commercial cultivation since they are high yielding and have beans weighing more than one gram.

Climate and Soil: Cocoa is a crop of humid tropics requiring well-distributed rainfall. A minimum of 90 to 100mm rainfall per month with an annual precipitation of 1500-2000mm is ideal. However, it can also be grown in other regions by supplementing rainfall with irrigation during dry periods. However, for successful cultivation the dry months should not exceed 3 to 4 months. This limits the distributions of cocoa to within 20^0 latitude on either side of the equator. Cocoa tolerates a minimum temperature of 15^{0}C and a maximum of 40^{0}C, but temperature around 25^{0}C is considered as optimum. It can be grown in place from sea level upto an elevation.

Cocoa grows on a wide range of soils but losse soils which allow root penetration and movement of air and moisture are ideal. It should retain moisture in the soil during dry season, as cocoa requires regular supply of moisture for proper growth. Though cocoa can be grown in soils with pH range from 4.5-8.0, it thrives better in neutral soil.

Propagation: Cocoa can be propagated from seeds or vegetatively from buds and cuttings. Seed pods may be collected from trees yielding 80 or more pods per year with pod weight 350-400g. Fresh beans from such pods should be used for sowing, as cocoa seeds lose their viability soon after they are taken out of pods. Before sowing, the seeds are rubbed with dry sand or wood ash to remove mucilage. The beans are planted with their pointed end upwards, either in plastic bags (25 x 15cm size) or in raised beds. If sown in beds, young seedlings are usually transplanted into polythene bags after about two weeks of germination. The seedlings are ready for transplantation to the field after about 3 to 4 months or they attain a height of 30 cm.

Cocoa can be also propagated vegetatively through cuttings, soft wood grafting, forkert method of budding etc., but there are limitations at present for adopting vegetative propagation on commercial scale.

Establishment of Plantation: Cocoa requires shade when young and also to a lesser extent when grown up. Young cocoa plants grow best with 50% full sunlight. Therefore, it is planted in arecanut and coconut gardens in our country or as a pure plantation in forestlands by thinning and regulating the shade suitably. It is planted at a distance of 2.5-3.0m both between and within rows, either in the beginning of the monsoon, in May-June or at the end of the southwest monsoon in September.

Cocoa under arecanuts and coconuts is the cropping systems, which can be adopted advantageously in Kerala, Karnataka and Tamil Nadu. In arecanut gardens where the spacing is 2.7 x 2.7m cocoa is interplanted in alternate rows at a spacing of 5.4 x 2.7m. in coconut gardens, it can be planted 2.7m apart in a single row. Under the double hedge system, cocoa is planted in two rows adopting a spacing of 2.7 m within the row and 2.5m between rows of coconut planted at a normal spacing of 7.5 x 7.5 and above.

Manures and Fertilisers: An annual application of 100g Nitrogen, 40 g Phosphorus and 140g Potash per tree per year in two split doses is recommended. During the first year of planting, the plants may be given one third of the above dose, while in the second and third year and above, two third and full dose of fertilizers may be applied uniformly around the base of the tree upto a radius of 75cm and forked and incorporated into the soil.

Irrigation: Cocoa plants require continuous supply of moisture for optimum growth and yield. During summer, the plants will have to be irrigated at weekly intervals. If adequate water supply is not ensured in summer months, the yield will be reduced and under mixed cropping systems, if there is severe drought, the yield of both the crops may be affected.

Pruning: The cocoa trees should be pruned regularly to develop a good shape. Cocoa grows in a series of storeys. The chupon or vertical growth of the seedlings terminates at the jorquette, where four or five fan branches develop. Further chupon develops just below the jorquette and continues its vertical growth till another jorquette forms and so on. When the first jorquette develops at a height of about 1.5m, the canopy will form at a height convenient for harvesting and other operations. Hence, all the fan branches arising from the main step are nipped off up to a height of about 1.0 to 1.5m or cut in the initial years of their growth. It is desirable to limit the height of the tree at that level by periodical removal of chupon growth. A second jorquette may be allowed to develop, if the first one formed was very close to the ground. Generally 3 to 5 fan branches are developed at each jorquette. When more fan branches develop, one or two weaker ones may be removed. The branches badly affected by pests and diseases also should be removed.

Harvesting: Cocoa flowers from the second year of planting and the pods take about 140 to 160 days to mature and ripen. Each pod will have 25 to 45 beans embedded in white pulp (mucilage). Generally cocoa gives two main crops in a year *i.e.*, September-January and April-June, off season crops may be seen almost all through the year, especially under irrigated condition.

Only ripe pods have to be harvested without damaging the flower cushion, at regular intervals of 10 to 15 days. The pods are harvested by cutting the stalk with the help of a knife. The harvested pods should be kept for a minimum period of two to three days before opening for fermentation. For breaking the pods crosswise, wooden billet may be used and the placenta should be removed together with husk and the beans and collected for fermentation. A pod will have about 30 to 45 seeds covered with pulp or mucilage.

Processing: Fermentation: The beans should be fermented to develop chocolate flavour, reduce bitterness, loose its viability, remove the mucilage coating and enable the cotyledons to expend. Fermentation is done immediately after collecting the beans from the pods. There are two popular methods of fermentation using either trays or boxes.

Box method: In this traditional method, boxes of various shapes and sizes are used. The smallest one has the measurements of 60cm x 60cm and will hold about 150kg wet beans. The bottom of the box has a number of holes of 1cm diameter spaced at about 10cm apart. Three such boxes are arranged in a row so those beans can be transferred from one box to the other. The beans are placed in the top most and covered with banana leaves or gunny bags. After 2 days, the bean should be uncovered and transferred into the second box and then to the third box after another 2 days. On the sixth day, fermentation is completed and beans can be taken out for drying.

Tray method: This method is used only for fermenting forestero cocoa beans. The normal size of the tray is 90 x 60 x 12cm with a capacity to hold about 40 kg wet beans. The bottom of the tray is either slotted or drilled to make 1cm holes on a 4cm sq. pattern. A minimum of 4 trays is needed for successful fermentation. All the trays are filled with beans. The topmost tray is covered with banana leaves or sacks. The fermentation is faster here and is completed in about 4 to 5 days. This method is more convenient for large growers as the trays can be easily handled and no mixing is required and the period of fermentation is less.

Basket method: Bamboo or cane basket of suitable size having one or two layers of banana leaves at bottom to drain the sweating is filled with the beans and the surface is covered with banana leaves. After one day the basket is covered with thick gunnysacks. The beans are mixed thoroughly on the third and fifth days and covered with gunnysacks. The fermentation will be completed at the end of the sixth day and the beans are withdrawn for drying.

Drying: After the fermentation, the beans can be dried by sun drying or artificial drying as the fermented cocoa beans have considerable moisture (55%). Sun drying is good as it gives superior quality produce when compared to artificial drying. The fermented beans are spread in a thin layer over a bamboo mat or cement floor and dried beans is around 6 to 7 percent.

During the monsoon period, artificial drying has to be adopted. Electric ovens or conventional Samoan type drier could be used. The duration of artificial drying varies from 48 to 72 hours at 60 to 70^{0}C. The drying of beans at high temperature should be avoided as it results in low quality end product. Slow drying in the initial stage has given better quality beans. Well-dried beans when shaken should give a metallic sound.

Grading and Storage: The flat, slaty, shrivelled, broken and other extraneous materials are removed. The cleaned beans are packed in fresh polythene-lined (150-200 gauge) gunny bags. The bags are kept on a raised platform of wooden planks. The beans should not be stored in rooms where spices, pesticides are fertilizers are stored as they may absorb the odour from these materials.

The roasted product of the dried beans is called as Cocoa, which are used for the manufacture of various products. When cacaonibs are ground, the resulting product is called Chocolate liquor or mass. The fat that is pressed from chocolate liquor is termed as cocoabutter. It is mainly used for the manufacture of chocolates, in pharmaceutical preparations and soap making.

Plant Protection

Many pests and diseases are known to infect cocoa and cause economic losses.

Sr. N.	Pest/disease	Symptoms	Control measures
1.	Mealy bugs (Planococcus lilacinus)	Adult and young ones suck tender shoots, cushions, flowers, cherelles and pods.	Spot application of monocrotophos 125ml. in 100 litres of water.
2.	Stem borer (Zeuzera coffeae)	Caterpillars bore into the branches and trunks.	Prune and destroy the affected branches, apply BHC paste.
3.	Aphids (Toxoptera aurantii)	Adults and nymphs feed on young leaves, succulent stem, flowers and small cherelles.	Spray dimethoate 1.5ml. In one litre water.
4.	Rodents –Rats and squirrels	Pods are damaged	Deep 10g of bromadiolone (0.005%) wax cakes on the branches for rots, keep machanical traps for squirrels.
5.	Blackpod disease (Phytophthora palmivora)	Pods turn chocolate brown to black, beans discoloured.	Remove infected pods at frequent interval, spray 1% Bordeaux mixture during monsoon twice at 45 days interval.
6.	Canker (P.palmivora)	Brownish water soaked lesions on the outer bark, which then turn into rusty deposits.	Remove the infected tissues and apply Bordeaux paste.

Cherelle Wilt: The shrivelling and mummifying of some young fruits is a common phenomenon in all coca gardens. In the early stages, the fruits lose their lusture and in four to seven days, the fruits shrivel. The fruits may wilt but do not abscise. Many factors are involved in the causation of this malady. The most important factors are insects, fungi, nutrient competition, over production etc. Hence, the remedial measures will depend upon the nature of the causative factors involved.

Plantation Crops

Arecanut

Arecanut palm *(Areca catechu L.)* is cultivated primarily for its kernel obtained from the fruit which is chewed in its tender, ripe or processed form. Arecanut belongs to the family Palmae. It is native of Malayan Archipelago, Philippines and other East Indian Islands. Commercial cultivation is confined only in India, Bangladesh and Sri Lanka. Kerala, Karnataka and Assam account for more than 90 per cent of the total area and production in our country. Arecanut production in India has now almost reached a level of self-sufficiency. Uses for Arecanut other than chewing are negligible. Its export prospects are also very much limited. Therefore, the present policy is not to expand the area under arecanut, but to adopt intensive cultivation and take up replanting of the aged and unproductive gardens. Inter and mixed cropping in arecanut gardens is advocated to augment the income from the existing arecanut garden. It is a monoecious palm and its inflorescence is spadix produced in the leaf axil and is completely enclosed in a sealed boat shaped spathe. The spadix is having a main rachis divided subsequently into secondary and tertiary rachis. It is a cross pollinated crop and fruit set normally varies form 12.0 to 40.0 per cent and the time taken from full bloom to maturity of the fruit ranges from 35 to 47 weeks.

Climate and Soil: The arecanut palm is capable of growing under a variety of climatic and soil conditions. It grows well from almost sea level upto an altitude of 1000 metres in areas receiving abundant and well distributed rainfall or under irrigated conditions. It is grown in soils such as laterite, red loam and alluvial soils. The soil should be deep and well drained.

Varieties: There are few local varieties known by the name of the place where they are grown and are furnished below:

Name of the local variety	Place where grown
South Kanara	Dakshina Kannada district and Kassargod district of Kerala
Thirthahalli	Malnad area of Karnataka
Sreevardhan	Coastal Maharashtra
Mettupalayam	Coimbatore District
Mahitnagar	West Bengal
Kahikuchi	Assam

Central Plantation Crops Research Institute, Regional Station, Vittal has released three improved cultivars, they are:

Name of the cultivar	Characters
Mangala	An introduction from China (VTL-3) early bearing, higher fruit set, higher yield (10 kg ripe nuts/palm/year), semitall variety
Subangla	A selection from Indonesia (VTL-11), yield 17.5 kg of nuts/palm at the age of 10 years.
Sreemangala	A selection from Singapore (VTL-170), yields 16.5 kg/palm at the 10th year.

Sreemangala A selection from Singapore (VTL-170), yields 16.5 kg/palm at the 10th year.

Selection and Raising of Planting Materials: Collection of seednuts should be confined to high yielding palms which commence to bear early as well as those which give more than 50 per cent of fruit set. From these selected mother palms, fully ripe nuts are alone collected. All undersized and malformed nuts must be rejected. Heavier seednuts (above 35 g) within a bunch are alone selected, as they give higher percentage of germination and produce seedlings of better vigour than lighter ones. The selected seednuts are sown immediately after harvest, 5 cm apart in sand beds under partial shade with their stalk ends pointing upwards. Sand is spread over the nuts just to cover them. The beds may be watered daily. Germination commence in about 40 days after sowing and the sprouts can be transplanted to the second nursery when they are about three months old. At this stage, the sprouts might have produced two to three leaves.

The Secondary nursery beds of 150cm width and of convenient length are prepared for transplanting the sprouts. The sprouts are transplanted at a spacing of 30cm x 30cm with the onset of monsoon. Partial shade to the seedlings can also be provided during summer by pandal or growing banana. Care should be taken to drain the nursery beds during the monsoon and to irrigate them during the dry months. Weeding and mulching should be done periodically. Seednuts can also be sown in polythene bags (25 x 15cm size, 150 gauge) after filling the bags with potting mixture containing 7 parts of loam or top soil, 3 parts of dried and powdered farm yard manure and 2 parts of sand. The seedlings will be ready for transplanting to the mainfield when they are 12 to 18 months old. Seedlings having 5 or more number of leaves should be selected. The height of seedlings and the time of planting has a negative correlation with the subsequent yield of the plant. Hence shorter seedlings with maximum number of leaves are remove with a ball of earth for transplanting. If the seedlings are raised in polythene bags, these can be straight way transported to any distance without much damage.

Planting: The planting is done during May –June with the onset of monsoon. Arecanut plants need adequate protection from exposure to the south western sun as they are susceptible to sun-scorch. Proper alignment of the palms in the plantation will minimise sun scorching of the stem. In the square system of planting at a spacing of 2.7m x 2.7m, the north south line should be deflected at angle of 35° towards west. The outermost row of plants on the southern and southwestern sides can be protected by covering the exposed to with areca leaves or leaf sheaths or by growing tall and quick growing shade trees.

Pits of 90 x 90 x 90cm are dug and the pils are filled with a mixture of top soil, powdered cowdung and sand to a height of 50 to 60cm from the bottom. The seedlings are planted in the centre of the pit, covered with soil to the collar level and pressed around. A shade crop of banana can be raised to give protection to the seedlings from sunscroch.

Manures and Fertilisers: Adequate supplies of plant nutrients in the soil throughout the life of the crop is essential to get high yield. Hence, an annual application of 100g Nitrogen, 40g Phosphorous and Potash in the form of fertilizer and 12kg each of green leaf and compost or cattle manure per bearing palm is recommended. Under rainfed conditions, half the quantity of fertilizers may be applied in April-May and the remaining quantity in September-October. Under irrigated condition, the first dose of fertilizer is applied only in February. Green leaf and compost can be applied in single dose in September-October. Irrespective of the age of the plants, full dose of green leaf and compost or cattle manure may be applied from the first year of planting itself while one-third of the recommended quantity of fertilizers in the first year, two-third in the second year and the full dose from the third year onwards. The first dose of fertilizers may be broadcast around the base of each plant after weeding and mixed with the soil by light forking, while the second dose is done in basins around the palm dug to a depth of 15 to 20cm and at 0.75 to 1m radius. In acidic soils, required quantity of the lime may be applied during the dry months and forked in.

Irrigation and Drainage: The palm should be irrigated once in four to seven days depending on the soil type and climatic factors. In Kerala, arecanut gardens are irrigated during dry months once in seven or eight days during November-December, once in the six days during January-February and once in 3-5 days during March, April and May. Adequate drainage should be provided during monsoon since the palms are unable to withstand waterlogging. Drainage channels should be 25 to 30 cm deeper than the bottom of the pits to drain excess water from the plot.

Harvesting and Processing: The stage of harvesting depends on the type of produce to be prepared for the market. The most popular trade type of arecanut is the dried, wholenut knows as chali or kottapak. Fully ripe, nine months old fruits having yellow to orange red colour is the best suited for the above purpose. Ripe fruits are dried in the sun for 35 to 40 days on drive levelled ground. For drying dehusking, sometimes fruits are cut longitudinally into two halves and sundried for about 10 days, the kernals are scooped out and given a final drying.

Another form of processing is by making Kalipak. The nuts of 6 to 7 months maturity with dark green colour are dehusked cut into pieces and boiled with water of dilute extract from previous boiling, a kali coating is given and dried finally. Kali is the concentrated extract obtained from boiling 3 to 4 batches of Kalipak. Many varieties of scented suparis are now prepared by blending the dried, broken bits of arecanut with flavour mixtures and packed.

Dehusking of arecanut is traditionally done by skilled manual labour with the help of tool which has a sickle shaped small pointed blade fixed on a plant. A simple device for dehusking arecanut, developed by CPCRI, Kasargod can also be used. The main advantage of this device is that any unskilled person can operate it. The outturn is about 60kg husked nuts in case of dried nuts and 30kg in case of green nuts if one person operators the device for a day of 8 hours.

Yield: More than 10 kg of ripe nuts per palm at the 10th year is considered as normal yield in any plantation.

Plant Protection: Important diseases and pests affecting arecanut are given below:

Pest/diseases	Symptoms	Control Measures
Pests		
Mites (Raoiella indica) (Oligonychus indicus)	Adults and young ones suck the lower surfaces of the leaves, causing them to turn yellow and bronzed in appearance	Spray the lower surface of leaves with dicofol 0.05%
Spindle bug (Calvalhoia arecae)	Adults and young ones suck the sap from the tender spindle resulting in loss of vigour and consequent death	Place 2 g of phorate granules taken in perforated polybags in the inner most leaf axils.
Inflorescence caterpillar (Tirathaba mundella)	Caterpillars feed on the flowers and clamp the inflorescence into a wet mass of frass with silky threads	Infected spadices may be forced open and sprayed with malathion 0.05%
Diseases		
Koleroga or mahili (Phytophthora arecae.)	Water-soaked lesions appearing on the nut surface near the perianth spread over the other parts giving the nut a dark green colour. Infected nuts shed without perianth.	Spray 1% Bordeaux mixture twice at 45 days interval during monsoon.
Bud rot (Phytopththora palmivora)	Affected spindle appear yellow, later changing to brown and finally the whole spindle rots.	Early removal of the infected tissue and treat the healthy tissue with Bordeaux paste. Drench the crown with 1% Bordeaux mixture as a prophylactic measure.
Anabe roga (Gonoderma luciderm)	Small brown irregular patches appear on the stem and a brownish exudate oozes out from these patches	Provide better drainage, isolate the affected palms by trenches, drench with 0.3% captain.
Yellow leaf disease (Mycoplasm like organisms)	Leaves become yellow, smaller, stiff and pointed, crown gets reduced, palm remains stunted with few or no nuts.	Regular manuring, ensure drainage, grow cover crops, remove the affected palms.

Plantation Crops

Cashew

Cashew is native of South Eastern Brazil, from where it was introduced to Malabar coast of India in the sixteenth century to cover bare hillside for soil conservation. It gained commercial importance in 1920. African countries produce large quantities of cashewnut, but they are not processed into consumable products because of difficulties in organising the native labour. India imports raw cashewnuts, which are processed and converted into cashew kernels and cashew shell liquid and re-exported to the countries, like USA, Canada, United Kingdom, Australia and Russia.

The cashew tree is a low spreading, evergreen tree with a number of primary and secondary branches and with a very prominent tap root and a well developed and extensive network of lateral and sinker roots. The fleshy peduncle the 'cashew apple' is juicy and sweet when ripe.

The apple varies in size, colour, juice content and taste. It is a rich source of vitamin C and sugar. The cashew fruit is a kidney shaped drupaceous nut, greenish grey in colour. The nuts vary in size, shape, weight (3 to 20g) and shelling percentage (15-30 per cent)

Climate and Soil: It is a hardy tropical plant and does not exact a very specific, climate. It can come up in places situated within 35 latitude on either side of the equator and also in the hill ranges upto 700 m MSL. It can grow well in places receiving rainfall from 50 cm to 250 cm and tolerate a temperature range of 25 to 49 C. it requires a bright weather and does not tolerate excessive shade.

Cashew is cultivated on a wide variety of soils in India like laterite, red and coastal sandy soil. To a limited extent, it is also grown on black soils. It can be also grown in hill slopes in virgin organic matter rich soils. They do not prefer water logged or saline soils.

Varieties: Since cashew is a highly cross-pollinated crop, planting of seedlings is not recommended now. Various cashew research centres have released improved clones, which are either selections from seedling population or hybrids. They are briefly described in the following table:

Name of the variety	Parentage	Remarks
A. Varieties released for cultivation in Andhra Pradesh		
BPP1	Hybrid tree No.1 x T.No. 273	13.2 % perfect flowers, apple medium sized, yellow, 60% juice content, 8 fruits / panicle, yield 17kg / tree, nut weight 5gm, shelling percentage 27.5%
BPP 2	Hybrid tree No.1 x T.No. 273	7% perfect flowers, 8 - 10 fruits per panicle, yield 19kg / tree, nut weight 4gm, shelling percentage 26.0%
BPP 3	Clonal selection from germplasm type	yield 16kg / tree, nut weight 6gm, shelling percentage 28.0%
BPP 4	Clonal selection from germplam type	yield 12.5kg / tree, nut weight 6gm, shelling percentage 23.0%
BPP 5	Clonal selection from T.No.1 (A.P.)	yield 42kg / tree, nut weight 5.2gm, shelling percentage 24.0%
BPP 6	Clonal selection from T.No.56 (A.P.)	yield 42kg / tree, nut weight 5.2gm, shelling percentage 24.0%
B. Varieties released for cultivation in Maharashtra		
Vengurla-1	Clonal selection from germplasm (Vengurla)	8% perfect flowers, intensive branching and compact type, yield 23kg / tree, nut weight 6.0gm, shelling percentage 31.0%
Vengurla-2	Clonal selection from Germplasm (West Bengal)	8% perfect flowers, short flowering and fruiting phase, yield 24kg / tree, nut weight 4.0gm, shelling percentage 32.0%
Vengurla-3	Ansur 1 X Vetore-56	25% perfect flowers, 7 fruits per panicle, yield 17kg / tree, nut weight 9.0gm, shelling percentage 27.0%
Vengurla-4	Midnapore Redx Vetore-56	35% perfect flowers, 6 fruits per panicle, yield 23kg / tree, nut weight 8.0gm, shelling percentage 31.0%
Vengurla-5	Ansur Early x Mysore Kotekar	50.5% perfect flowers, 3-4 fruits per panicle, yield 21kg / tree, nut weight 4.5gm, shelling percentage 30.0%

Propagation

- Seed Propagation: Seed propagation is seldom practised now except to raise the rootstock materials. Seeds should be collected during the month of March to May and the heavy seed nuts, which sink in water, are alone mixed with 2 parts of fine sand. They take normally 15 to 20 days for germination.
- Vegetative propagation: One year old shoots as well as current season shoots are used for air-layering. Though there is good root development in this method, heavy mortality occurs both at nursery stage and on the main field. Other drawbacks of this method are that it is cumbersome, time consuming and production of a limited number of layers per tree and as such layers are also not suitable for cyclone prone areas as they do not have a tap root providing less anchorage to soil.

The vegetative propagation through cuttings is seldom practised, as success is very less. Similarly veneer grafting, side grafting and patch budding are also reported to be successful but the nursery period is quite long, 3 to 4 years. Recently 'epicotyl grafting' and 'soft wood grafting' are recommended for commercial scale adoption. In the case of epicotyl grafting, tender seedlings with height of 15 cm are selected as root stocks and a 'V' shaped cut is made after beheading it at a height of 4 to 6 cm from the cotyledons connective. The precured scion is collected and a wedge is made at the base of it, so as to exactly fit in the cut made in the stock. The scion is exactly fitted in the stock and tied with polythene strips. The success of epicotyl grafting varies from 50 to 60 per cent and depends upon high humidity, temperature, freedom from fungal disease, number of rainy days and rate of cambial growth. When the above method is adopted in 30 to 40 days old seedling, it is known as soft wood grafting. The success varies from 40 to 50 per cent.

Planting: Pits of 45 x 45 x 45 cm are dug and filled with a mixture of top soil, 10 kg of farm yard manure and one kg of neem cake at a distance of 7m x 7m either way during June July and planted. In the case of seedling, 45 days old seedlings are transplanted.

Training and Pruning: All the side shoots must be removed upto a height of atleast 2m from the ground to cause the branches to form and spread out from the upper section of the trunk. Periodical pruning of dead wood and criss cross branches during the month of July is recommended to minimise the losses through diseases such as dieback and to increase the yield.

Manures and Fertilisers: The recommended schedule is

Age of the plant	Manures and fertilisers per tree			
	FYM or Compost (kg)	Nitrogen (g)	Phosphorus (g)	Potash (g)
One year old	10	50	25	25
Two year old	20	100	50	50
Three year old	20	150	75	75
Four year old	30	150	75	75
Five year old and above	50	500	125	125

Top Working: As most of the existing cashew plantations are of seedling progenies, the yield level is very low and highly erratic. Hence, top working with improved clones are suggested now. Trees of 20 to 25 years old are beheaded at a height of 0.5 m from the ground during December-February. A paste, made using 50 g, each of BHC 50 per cent wettable powder and copper oxychloride in a litre of water, should be applied all over the stump to check any infection by invading pathogens and borer insects. Profuse sprouting normally results in but only 10 to 15 healthy shoots and properly spaced on the stumps are alone retained. These shoots are grafted at softwood stage (cleft grafting) when they are about 40 to 50 days old. 7-8 successful grafts may be encouraged to grown and the sprouts should be periodically removed. Top worked trees grown vigorously due to the well established root system and they start yielding about 4kg per tree from the second year of rejuvenation and the yield gradually increases to stabilise at 8 kg from the fourth year of top working.

Yield: The yield depends upon many factors. Individual trees, which yield more than 6 kg after 15 years, are considered as good yielders.

Cashew Processing: Processing consists of roasting, shelling, extracting the oil, peeling, grading and packing

a. *Roasting:* Roasting makes the shells brittle, besides making the extraction of kernels easier. A slight underoasting or overroasting adversely affects the quality and recovery of kernels. In the open pan roasting method, one kilogram of nuts are kept in shallow iron pans or earthen pots and are heated over an open fire. The nuts are rapidly turned to prevent charring. The roasted nuts are then removed from the pan and thrown on the floor. They are quickly covered with earth, which would absorb shell oil adhering to the roasting nuts and also cool them. The nuts are then subjected to subsequent operations. The cashew nuts are also roasted by drying under sun for two to three days when they lose much of the moisture contents and become brittle enough for shelling. The other improved methods of roasting cashewnuts are

i. *Continuous Roasting Process:* The principle adopted in this system is the same as in the case of open pan roasting method. This plant consists of a single walled or double walled rotating metallic drum. In the case of double walled drum, the smoke or gases produced during roasting escape through the interspace between the two walls of the drum and are condensed to shell oil by a cooling system but in the case of single walled drum, the gases that escape from the nuts during the process of roasting are allowed to escape through a chimney provided at the lower end and there is no provision for collecting shell liquid.

ii. *Oil bath Process:* In this method, the nuts are held in wire trays and are passed through a bath of cashew shell oil maintained at a temperature of 200 to 202^{0}C for a period of three minutes whereby the shell oil is recovered from the shells to the maximum possible extent. This process ensures uniform roasting of nuts and eliminates charring of kernels.

b. *Shelling:* After roasting, shelling is done by labour. Each nut is placed edgewise and cracked open with a light wooden mallet and the kernel extracted with or without the help of a wire prong. Care has to be taken that the inner kernel is intact and is not broken into bits.

c. *Peeling:* Removal of a thin outer brown skin is done by hand with the help of a safety pin or small hand knife. Peeling is made easier when the kernels are subjected to a heat treatment for about four hours in a drying chamber. After peeling, the kernels are spread out indoors on cement flooring so that they may absorb some moisture and become less brittle. This prevents the tendency to break easily during grading.

d. *Grading:* Grading is done based on "counts" or number of kernels per pound. The kernels, which have no split, are separated as 'wholes'. These are again separated into six grades as 210,240,280,320,400 and 430 whole nuts per pound. The graded kernels should be fully developed, ivory white in colour and free from insect damage and black or brown spots. The broken and split kernels are then separated and classified as standard and scorched pieces, splits, butts, small pieces and each grade is separately packed.

e. *Packing:* Packing is done in this. In this method, the air inside the tine is exhausted and they are recharged with Co_2 before they are sealed air-tight.

Plant Protection

The following two pests are economically important in cashew plantations.

1. *Stemborer:* The grub bores into the trunk and the roots. Swabing the trunk with BHC 50% and drenching the soil around the base of the tree with BHC 50% are recommended.
2. *Tea Mosquito Bug:* Adults and nymphs suck the sap from the tender plant parts. Spraying endosulfan 0.05% thrice, first at the time of emergence of new flushes, the second at floral formation and third at fruit set is recommended.
3. *Die Back or Pink Disease:* Spraying any copper fungicides besides pruning the dead twigs are suggested.

Plantation Crops—Coffee

Coffee, native of Ethiopia, was introduced into India sometime during 1600 AD by a Muslim pilgrim, BabaBudan on the hills near Chikmangalur. Now coffee cultivation is mainly confined to the State of Karnataka, Kerala, Tamilnadu and Andhra Pradesh and on a limited scale to Arunanchal Pradesh, Assam, Madhya Pradesh, Manipur, Meghalaya, Mizoram, Nagaland, Orissa, Sikkim, Tripura and West Bengal. Coffee belongs to the family Rubiaceae. Though the genus Coffee consists of about 70 species, only three species are of economic importance. They are (i) C. arabica (Arabica coffee) (ii) C. cenephora (Robusta coffee) and (iii) C. Liberia (Tree coffee). The first two species are extensively cultivated.

Climate and Soil: Climatological factor like rainfall, temperature, elevation and aspect can influence economic production of coffee much more than soil factors. Soil should be deep, well drained, slightly acidic in reaction and rich in organic matter content. The optimum soil and climatic requirements for arabica and robusta under south India conditions are as follows:

S. N.	Name	Requirements
1.	Coffee Arabica	Elevation – 1000-1500m MSL Annual rainfall – 1600-2500 mm Shade – Needs medium to light shade Temperature – 15-25^{0}C Relative humidity- 70-80% Soil – Deep friable, porous, rich in organic matter moisture retentive, slightly acidic pH 6.0 to 6.5
2.	Coffee Robusta	Elevation – 500-1000m MSL Annual rainfall – 1000-2000 mm Shade – Needs uniform thin shade Temperature – 20-30^{0}C Relative humidity- 80-90% Soil – Deep friable, porous, rich in organic matter moisture retentive, slightly acidic pH 6.0 to 6.5

Coffee cultivation is confined mostly to the hilly tracts of Western and Eastern Ghats. A well distributed rainfall is preferable for coffee with a dry months from December-March. Summer showers are important for flowering and failure of blossoms showers leads to crop loss.

Varieties: Crop improvement work carried out at Central Coffee Research Institute, Balahanur, Karnataka has resulted in the release of a number of superior selections in Arabica coffee and the popular once are furnished below.

S. N.	Name of the variety	Parentage	Characters
1.	S.795 (Sln.3)	A cross between S.288 x Kent	Resistant to leaf rust race 1 and 11, popular among growers.
2.	Sln.7 (San Ramon hybrids)	San Ramon-a short internode arabica type	Dwarf in stature, but segregates to tall by 30% also.
3.	Sln.8 (Hibrido-de -Timor)	A spontaneous hybrid of robusta-arabica	Highest vertical resistance to leaf rust, phenotype and bean quality resembles arabica.
4.	Sln.9	Sln. 8 x Tafarikela	Drought hardy, suitable to different coffee zones.
5.	Sln.10 (Catura crosses)	Caturra (dwarf arabica coffee, Portugal) x S.795 or Sln.8	Drought hardy, suitable to different coffee zones.
6.	Cauvery	F4 of a cross between Caturra x Hibrido-de-Timor	Comes to bearing within three years yield potential 2.5 t/ha, amenable for close planting, resistant to almost all races of leaf rust.

Nursery Management

Healthy and mature fruits of normal size and appearance, three quarters to fully ripe are harvested from specially selected and marketed coffee plants for use as seed bearers. Floats are discarded, the sound fruits are pulped, the beans drained and sieved to remove defective beans. The beans are then mixed with sieved wood-ash, evenly spread out to a thickness of about 5cm and allowed to dry to facilitate uniform drying. Excess ash is rubbed-off after five days of drying. Germination beds, raised to a height of about 15 cm, one metre width and of convenient length are prepared. Four baskets of fully mature cattle manure or compost, about 2 kg. of finely sieved agricultural lime and 400g of rock phosphate are incorporated in a bed measuring 1m x 6m.

Sowing: Seeds should be sown with the flat side facing the soil at a distance of 1.5– 2.5cm from one another in regular rows. A thin layer of fine soil is then spread. The bed is covered with a layer of about 5cm of paddy straw. The beds are watered daily and protected from direct sunlight by an overhead pandal. The seeds germinate in about 45 days. The seedlings are then transplanted to secondary nursery beds or raising polybag plants.

Transplanting in Bags: Coffee seedlings are transplanted to polythene bags of 23 cm x 5 cm with 150 gauge thick in February or March when they are at the button or topee stage. The bags are filled with a prepared mixture of 6 parts of jungle soil, 2 parts of well rotten sieved cattle manure and 1 part of fine sand. At the time of transplanting it is preferable to slightly nip the tap root of the seedling. Transplanting is done preferably in the early morning hours or late in the afternoon. Regular watering and after-care of the seedlings should follow. Seedlings may be manured once in 2 months with urea dissolved in water, 20g. urea in 4.5 liters of water is sufficient for an area of 1 square metre. Adequate protection is given against nursery diseases and pests. Overhead shade in the nursery has to be thinned and finally removed after the onset of monsoon.

Preparation of Land: Clean felling is not advocated. Selective retention of desired species of wild shade trees is essential. The land should be divided into blocks of convenient size with foot path and roads laid out in-between. In steapy area, terracing and contour planting may also be adopted.

Spacing for arabica and robusta coffee is 2 to 2.5m and 2.5 to 4m respectively on either way. A close planting at 1-1.5 m eitherway and reduce the population by half after one or two harvests is good. Pits of 45 cm, are usually opened after the first few summer shower and seedlings of 16 to 18 months old are planted during June or September–October. A hole is made in the centre of the pit after levelling the soil. The seedling is placed in the hole with its tap root and lateral roots spread out in proper position. The hole is then filled. The soil around the seedlings is packed firmly and evenly in such a way that 3cm high above the ground to prevent stagnation of water around the collar. The seedlings are provided with cross stakes to prevent wind damage.

Training and Pruning: Training of the bush is necessary to have a strong frame work which promotes producing of bearing wood. Coffee is trained in two systems *viz.*

1. *Single Stem System:* When the plant reaches a height of 75 cm in Arabica or 110 to 120 cm in robusta, lit is topped. This helps to restrict vertical growth, facilitate lateral spreading and increase the bearing area. In this system, a second tier is also allowed sometimes depending upon the soil fertility and plant's vigour. (i) Multiple stem system–It is common in Kenya, Tanzania, but not practiced in India.

Pruning in coffee is generally done immediately after harvest and till the onset of monsoon. It is essentially a thinning process and is done mainly to divert the vigour of the plants to certain parts by pruning the other parts. Pruning involves (a) centering-removal of the vegetative growth upto 15 cm redius from the centre and upto the first node of all primary branches (b) desuckering – removal of small sprouts arising from the axis of the leaves which otherwise grow towards the inner side and cause shade and become unproductive wood and (d) nipping-growing tip of primary branches is removed to encourage secondaries and tertiaries.

Soil Management: Soil management practices aim at conserving soil and water and in general to make the soil perform its functions satisfactorily. It includes the following practices in coffee.

a. *Digging:* In the new clearing, the field is given a thorough digging to a depth of about 35 to 45 cm towards the end of the monsoon. All weeds and vegetative debris are completely turned under and buried in the soil while the stumps are removed. Once the coffee plants have closed in, annual digging is not done.

b. *Scuffing or Soil Stirring:* In established coffee fields, scuffling or soil stirring is done towards the beginning of the dry period. It controls weeds and also conserves soil moisture.

c. *Trenching:* Trenches and pits are dug or renovated in a staggered manner between rows of coffee along the contour during August-October when the soil is fairly easy to work. These are 50 cm wide and 25cm deep and can be of any convenient length.

d. *Mulching:* Mulching young coffee clearings helps to maintain optimum soil temperature and conserve soil moisture and acts as an effective erosion control measure. Mulching also adds to fertility of the soil.

e. *Weed Control:* New clearings are hand-weeded three to four times a year and established coffee two to three times. During the monsoon, the weeds are slashed back. Another weeding is done towards the end of the monsoon. Clean weeding is generally done during the post-monsoon period. Chemical weedicides have gained popularly in larger plantations. Gramoxone at 1.25 litres in 450 litres of water per hectare has been found to be the best. This should follow slash weeded plots after 10-15 days.

f. *Irrigation:* Sprinkler irrigation is mainly used as an insurance against failure of good blossom or backing showers. It is also used on young plantations, marginal areas where water is available in plenty to help in establishment of coffee and shade.

g. *Soil Acidity and Liming:* The heavy rainfall in coffee growing zones of South India brings about leaching in calcium and magnesium leading to soil acidity. Besides, continuous use of acid forming fertilizers like ammonium sulphate also make the soil acidic. As the ill effects of soil acidity are more, periodical lime application is essential to correct the soil pH for good productivity. The quantity of lime to be applied is based on soil pH and lime requirement. Agricultural lime and dolomitic lime are the most commonly used liming materials. It can be applied to the soil at any time during the year provided there is a gap of one month or a few showers between lime and fertilizer applications. It is also desirable to apply lime when there is sufficient moisture in the soil for quick response.

Shade Management

Under the climatic conditions existing in India, Coffee is being cultivated under shade. It comprises of two canopies-lower or temporary and upper or permanent. High light intensities and temperature prevailing during the drought period are not conducive for normal and healthy growth of coffee plants. Therefore, there is necessity for protecting the coffee plants during the above period by providing both temporary (lower) and permanent (upper) shade trees.

Dadap (Erythrina lithosperma) is used as a lower canopy shade in India. It is planted along with coffee in new clearings. When stakes are planted in June, they grow quickly using the moisture available in the soil. Next to dadap, silver oak (Grevillea robusta) is the most commonly used tree for temporary shade. The most popular permanent shade trees found in South India are Albizzia lebbek, A.moluccana, A.odoratissima, Artocarpus integrifoila, Cedrella toona, Dalbergia latifolia, Ficus glomerata, F. infectoria, F. retusa and Maesopsis eminii.

Permanent shade trees are generally planted about 12 to 14 metres apart. It is advisable to plant a large number initially and thin out as the trees grow and spread out. The trees have to be regulated in such a way that in course of tome, they have their canopy about 10 to 14 metres above the coffee. Shade trees require constant attention by way of pruning and lopping to provide the required filtered shade to coffee. The most convenient time to regulate shade is after pruning and liming.

Manuring: Coffee plants produce every year fresh wood for the succeeding crop concomitant with the function of maturing the current betties. Hence, they require a regular supply of nutrients. Besides, being grown in heavy rainfall area, the losses of nutrients due to leaching and fixation are to be offset by regular application of adequate quantities of fertilizers. The peak periods of demands for nutrients are at the time of flowering fruit-set and development and maturation of the crop. Based on all these factors, Coffee Board recommends the following dose of fertilizer for coffee.

Manurial Recommendation for Coffee

	Pre-blossom March N : P_2O_5 : K_2O	Post-blossom May N : P_2O_5 : K_2O	mid-monsoon August N : P_2O_5 : K_2O	Post-monsoon October N : P_2O_5 : K_2O	Total
ARABICA					
Young coffee 1st year after planting	15:10:15	15:10:15	-	15:10:15	45:30:45
2nd and 3rd year	20:15:20	20:15:20	-	20:15:20	60:45:60
4th year	30:20:30		-	30:20:30	80:60:80
Bearing coffee 5 years and above for less than one tonne/ha. Crop	40:30:40	40:30:40	20:0:0	40:30:40	140:90:120
For 1 tonne/ha. and above	40:30:40	40:30:40	40:30:40	40:30:40	160:120:160
ROBUSTA					
For less than 1 tonne/ha. Crop	40:30:40	-	-	40:30:40	80:60:80
For 1 tonne/ha. And above	40:30:40	40:30:40	-	40:30:40	120:90:120

The leaf mulch beneath the coffee is swept towards the base and the fertilizers are applied in a broad circular band about 30 cm away from the stem. They are then incorporated into the soil, with a fork or stick and covered by mulch. As a supplement to soil applications of fertilizers, foilar spraying with (Urea 0.5 kg, ammophos (20:20) 0.5kg and muriate of potash 350g dissolved in 200 litres of water or Bordeaux mixture may be given) during periods of slow growth, flowering and fruit-setting.

Harvesting and Processing

Coffee fruits should be picked as and when they become ripe to get better quality. Arabica comes for harvesting earlier since they take 8-9 months for fruit development from flowering while robusta takes 10-11 months. Picking is done by hand. The first picking consists of selective picking of ripe berries often seen in the outer portion of the node and is called fly picking. Thereafter, there will be 4-6 main picking at 10-15 days intervals and final harvest *i.e.* stripping consists of picking of still remaining green berries on the plant. Coffee is processed in two ways, (a) Wet processing to prepare plantation or parchment coffee and (b) dry method by which cherry coffee is prepared.

Preparation of parchment coffee:

i. *Pulping:* This method requires pulping equipment and adequate supply of clean water. Fruits should be pulped on the same day to avoid fermentation before pulping. Fruits may be fed to the pulper through siphon arrangements to ensure uniform feeding and to separate lights and floats from sound fruits. The pulped parchment should be sieved to eliminate any unpulped fruits and fruit skin. The skins separated by pulping should be let away from the vats into collection pits so that microbial decomposition of the skin will not affect the bean quality when it gets mixed up with the bean.

ii. *Demucilaging and Washing:* The mucilage on the parchment skin can be removed by

   a. *Natural Fermentation:* The mucilage breaks down in the process of fermentation and it take 24 to 36 hours for arabica and 72 hours for robusta. Cool weather delays the process of fermentation. Under fermented or over fermented beans affect quality. When correctly fermented the mucilage comes off easily and the parchment does not stick to the hand after washing and the beans feel rough and grittly when squeezed by hand. When the mucilage breakdown is complete, clean water is let in and the parchment washed pebble clean with three to four changes of water.

   b. *Treatment with Alkali:* Removal of mucilage by treatment with alkali takes about one hour for arabica and one and a half to two hours for robusta. The beans obtained after pulping are drained off excess water and spread out in the vats uniformly and furrowed with wooden ladles with long handles. A 10% solution of caustic soda (sodium hydroxide) is evenly applied into the furrows using a water can. 10 litres of the alkali is sufficient to treat 25 to 30 forlits (one forlit = 40 litres) of parchment. The parchment is agitated thoroughly by the ladles so as to make the alkali to come into contact with the parchment and trampled by feet for about half an hour. When the parchment is no longer slimy and makes a rattling noise, clean water is let in and the parchment washed clean with three or four changes of water.

c. *Removal of Mucilage by Friction:* There are machines, which pulp and demucilage the beans in one operation. However, a number of naked and bruised beans may result in the parchment. It is, therefore, necessary to adjust the machines carefully to obtain uniform pulping and demucilaging.

The parchment is spread on clean tiled or concrete drying floor to be dried slowly by spreading to a thickness of about 7 to 10 cm. Stirring and turning over coffee, at least once an hour, is necessary to facilitate uniform drying. The parchment should be heaped up and covered in the evening until next morning. Sun drying may take about 7 to 10 days under bright weather conditions. At the right stage of dryness the parchment becomes crumbly and the beans split clean without a white fracture when bitten between the teeth. Drying is complete when a sample forlit of coffee records the same weight for two days consecutively. At this stage, coffee is shifted to the stores and bagged in clean, new gunnies.

When coffee is being dried, all naked beans, pulper nipped and bruised beans, blacks, greens and other defective beans are sorted out and dispatched to curing works separately.

Preparation of Cherry: For preparation of cherry coffee, fruits should be picked as and when they ripe. Greens and under-ripe fruits should be sorted out and dried separately. The fruits should be spread evenly to a thickness of about 8 cm on clean drying ground in which the cherries are stirred and ridged atleast once every hour. The cherry is dry when a fistful of the drying cherry produces a rattling sound when shaken and a sample forlit records the same weight on two consecutive days. The cherry should be fully dry at the end of 12 to 15 days under bright weather conditions.

Plant Protection

The important pest and diseases in coffee, their symptoms/ damage and the control measures are presented below.

Pest/diseases	Symptoms	Control measures
Pests		
1. Mealy bug (Planococcus lilacinus and P.citri)	Serious foilar parasites, attack starts from a few isolated bushes and then spreads to others.	Prune the affected bushes, spray with foilthion 50EC@ 300ml or lebaycid 1000@150 in 200 litres of water.
2. Green bug (Coccusviridis)		
3. White stem borer (Xylotrechus quadripes)	Plants show unhealthy signs like wilting and yellowing of leaves.	Provide good shade, burn the infested plants in situ and swap with BHC 50Wp @ 2kg in 100 litres of water, padding with monocrotophos (1:2)
4. Short-hole borer (Xylosandrus compactus)	Attacked plants dry up, extensive tunnelling within the branches seen	Prune and burn the affected branches
5. Cockchafers (Holotrichia spp.)	Grubs feed on feeder roots of coffee, old plants withstand but young ones often wilt and die.	Drench the soil around the base of the plants with lindane 20 EC @ 35 ml in 10 litres of water.
6. Nematodes (Pratylenches coffeae)	General stunting, Yellowing of leaves, distortion of roots leading to die-back, affected plants easily get dislodged due to poor anchorage.	Remove the affected plants, expose the site to the sum, plant robusta seedlings during next season.
Diseases		
1. Leaf rust (Hemileia vastaririx)	Pale yellow spots on the lower surface of leaves, later turn to orange yellow powdery mass, infected plant exhibit defoliation	Spray Bordeaux mixture (0.5%) four time a year pre blossom, pre monsoon, mid monsoon, and post monsoon months
2. Blackrot (Koleroga noxia)	Blackening and rotting of the affected leaves, twigs and developing berries	Proper shade regulation, centering and handling the affected bushes to prevent secondary spread, spraying with 1% Bordeaux mixture
3. Brown blight. twig blight dieback (Collectorichum gloeosporioides)	Small water soaked lesions on margins of leaves and slowly extend causing drying of margins, often bushes start drying downwards with shedding of berries	Prune badly affected plants during dry months, spray 0.5% Bordeaux mixture
4. Root diseases (Fomes noxius, Poria hypolateritia, Rosellinia bunodes)	Affected plants show gradual yellowing of leaves, defoliation followed by death of above ground parts	Uproot the affected plant and burn, dig trenches of 60 cm deep and 30cm width to isolate the affected bushes, keep fallow for 6 months, apply organic manures 10-15 kg per pit.

Coconut

The Coconut Palm (*Cocos nucifera*) is a member of the Family Arecaceae (palm family). It is the only species in the genus *Cocos*, and is a large palm, growing to 30 m tall, with pinnate leaves 4-6 m long, pinnae 60-90 cm long; old leaves break away cleanly leaving the trunk smooth. The term *coconut* refers to the nut of the coconut palm, which commonly is referred to as a fruit, but is not.

The coconut palm is grown throughout the tropical world. It is so common in Kerala that one of its popular name in Malayalam is Keram derived from Kerala. Infact it is the *state tree* of Kerala. Another common name of coconut in Malayalam is Thenga and the palm itself is called Thengu. Virtually every part of the coconut palm has some human use, culinary and non-culinary and is justifiably called as a kalpa vruksham.

Origins and Cultivation

The origins of this plant are the subject of controversy, with some authorities claiming it is native to south Asia, while others claim its origin is in northwestern South America. Fossil records from New Zealand indicate that small, coconut-like plants grew there as long as 15 million years ago. Even older fossils have been uncovered in Rajasthan, Tamil Nadu, Kerala and Maharashtra, India. Regardless of its origin, the coconut has spread across much of the tropics, probably aided in many cases by sea-faring peoples. The nut is light and buoyant and presumably spread significant distances by marine currents. Nuts collected from the sea as far north as Norway have been found to be viable (and subsequently germinated under the right conditions). In the Hawaiian Islands, the coconut is regarded as a Polynesian introduction, first brought to the islands by early Polynesian voyagers from their homelands in the South Pacific. They are now ubiquitous to most of the planet between 26°N and 26°S.

The coconut palm thrives on sandy soils and is highly tolerant of salinity. It prefers areas with abundant sunlight and regular rainfall (750 to 2,000 mm annually), which makes colonizing shorelines of the tropics relatively straightforward.

Coconuts also need high humidity (70–80%+) for optimum growth, which is why they are rarely seen in areas with low humidity (*e.g.* the Mediterranean), even where temperatures are high enough (regularly above 24°C). They are very hard to establish in dry climates and cannot grow there without frequent irrigation. They may grow but not nut properly in areas where there is not sufficient warmth, like Bermuda.

Coconut palms are intolerant of freezing weather. They will show leaf injury below 34ºF (1ºC), defoliate at 30ºF (-1ºC) and die at 27ºF (-3ºC). There are rare reports of coconut palms surviving (with severe damage) to 20ºF (-7ºC). One night of freezing weather can set the growth of a coconut palm back about 6 months.

The flowers of the coconut palm are polygamomonoecious, with both male and female flowers in the same inflorescence. Flowering occurs continuously, with female flowers producing seeds. Coconut palms are believed to be largely cross-pollinated, although some dwarf varieties are self-pollinating.

Growing in the United States

The only two states in the U.S. where coconut palms can be grown and reproduce outdoors without irrigation are Hawaii and Florida. Coconut palms will grow from Bradenton southwards on Florida's west coast and Melbourne southwards on Florida's east coast. The occasional coconut palm is seen north of these areas in favoured microclimates in the Tampa-St. Petersburg-Clearwater metro area and around Cape Canaveral. They may likewise be grown in favoured microclimates on the barrier islands near the Brownsville, Texas area. They may reach fruiting maturity, but are damaged or killed by the occasional winter freezes in these areas. While coconut palms flourish in south Florida, unusually bitter cold snaps can kill or injure coconut palms there as well. Only the Florida Keys provide a safe haven from the cold as far as growing coconut palms on the U.S. mainland.

The farthest north in the United States a coconut palm has been known to grow outdoors is in Newport Beach, California along the Pacific Coast Highway. In order for coconut palms to survive in Southern California they need sandy soil and minimal water in the winter to prevent root rot, and would benefit from root heating coils.

Pests and Diseases

Coconuts are susceptible to the phytoplasma disease lethal yellowing. One recently selected cultivar, 'Maypan', has been bred for resistance to this disease. The fruit may also be damaged by eriophyid mites.

The coconut is also used as a food plant by the larvae of many Lepidoptera species, including the following *Batrachedra spp*: *B. arenosella*, *B. atriloqua* (feeds exclusively on *Cocos nucifera*), *B. mathesoni* (feeds exclusively on *Cocos nucifera*), and *B. nuciferae*.

The nut

Botanically, a coconut is a *simple dry nut* known as a fibrous drupe (not a true nut). The husk (mesocarp) is composed of fibres called *coir* and there is an inner "stone" (the endocarp). This hard endocarp (the outside of the coconut as sold in the shops of non-tropical countries) has three germination pores that are clearly visible on the outside surface once the husk is removed. It is through one of these that the radicle emerges when the embryo germinates. Adhering to the inside wall of the endocarp is the *testa*, with a thick albuminous endosperm (the coconut "meat"), the white and fleshy edible part of the seed. The endosperm surrounds a hollow interior space, filled with air and often a liquid referred to as coconut water, not to be confused with coconut milk. Coconut milk is made by grating the endosperm and mixing it with (warm) water. This produces a thick, white liquid called coconut milk that is used in much Asian cooking, for example, in curries. Coconut water from the unripe coconut, on the other hand, is drunk fresh as a refreshing drink.

When viewed on end, the endocarp and germination pores gives to the fruit the appearance of a *coco* (also Côca), a Portuguese word for a scary witch from Portuguese folklore, that used to be represented as a carved vegetable lantern, hence the name of the nut. The specific name *nucifera* is Latin for *nut-bearing*. When the coconut is still green, the endosperm inside is thin and tender, often eaten as a snack. But the main reason to pick the nut at this stage is to drink its water; a big nut contains up to one liter. When the nut has ripened and the outer husk has turned brown, a few months later, it will fall from the palm of its own accord. At that time the endosperm has thickened and hardened, while the coconut water has become somewhat bitter.

When the nut is still green the husk is very hard, but green nuts rarely fall, only when they have been attacked by moulds, etc. By the time the nut naturally falls, the husk has become brown, the coir has become dryer and softer, and the nut is less likely to cause damage when it drops. Still, there have been instances of coconuts falling from palms and injuring people, and claims of some fatalities. This was the subject of a paper published in 1984 that won the Ig Nobel Prize in 2001. Falling coconut deaths are often used as a comparison to shark attacks; the claim is often made that a person is more likely to be killed by a falling coconut than by a shark. However, there is no evidence of people being killed in this manner. However William Wyatt Gill, an early LMS missionary on Mangaia recorded a story in which Kaiara, the concubine of King Tetui, was killed by a falling green nut. The offending palm was immediately cut down. This was around 1777, the time of Captain Cook's visit.

In some parts of the world, trained pig-tailed macaques are used to harvest coconuts. Training schools for pig-tailed macaques still exist in southern Thailand and in the Malaysian state of Kelantan. Competitions are held each year to discover the fastest harvester.

Opening a Coconut

To open a coconut, pierce the softest "eye" with a skewer and drain the water. Then strike the coconut against a hard surface (such as concrete or a kitchen surface). It should break open similarly to an egg, cracking in more than one place. However, quite a lot of force is required. An easier way is to drain the water, then wrap the coconut in a towel and hit it with a hammer. A way to open a fresh coconut is to take a long, heavy, knife (such as a machete) and score a line across the middle of the coconut by striking repeatedly then rotating. The final stroke should be heavier than the previous to crack the coconut along the scored line. This is the normal method in countries with plentiful coconuts.

A coconut is husked with the use of a sturdy, pointed metal stake set about waist high in concrete. The coconut is struck onto the point nearer the pointed end and twisted. The process is repeated until the husk is loose and removed by hand. An experienced husker can husk a coconut in under ten seconds.

A slower method requires only a sharp rock to beat the husk loose, pulling off fibres by hand. With this method a coconut can be husked in five to ten minutes.

Uses

Nearly all parts of the coconut palm are useful, and the palms have a comparatively high yield (up to 75 fruits per year); it therefore has significant economic value. The name for the coconut palm in Sanskrit is *kalpa vruksham*, which translates as "the tree which provides all the necessities of life". In Malay, the coconut is known as *pokok seribu guna*, "the tree of a thousand uses". In the Philippines, the coconut is commonly given the title "Tree of Life".

Uses of the various parts of the palm include:

Culinary

- The white, fleshy part of the seed is edible and used fresh or dried in cooking.
- The cavity is filled with coconut water which contains sugar, fibre, proteins, anti-oxidants, vitamins and minerals. Coconut water provides an isotonic electrolyte balance, and is a highly nutritious food source. It is used as a refreshing drink throughout the humid tropics. It can also be used to make the gelatinous dessert nata de coco. Mature fruits have significantly less liquid than young immature coconuts; barring spoilage, coconut water is sterile until opened.

- Sport fruits are also harvested, primarily in the Philippines, where they are known as *macapuno.*
- Coconut milk is made by processing grated coconut with hot water or milk, which extracts the oil and aromatic compounds. It should not be confused with the coconut water discussed above, and has a fat content of approximately 17%. When refrigerated and left to set, coconut cream will rise to the top and separate out the milk.
- The leftover fibre from coconut milk production is used as livestock feed.
- The sap derived from incising the flower clusters of the coconut is fermented to produce palm wine, also known as "toddy" or, in the Philippines, *tuba.* The sap can also be reduced by boiling to create a sweet syrup or candy.
- Apical buds of adult plants are edible and are known as "palm-cabbage" (though harvest of these kills the palm).
- The interior of the growing tip may be harvested as heart-of-palm and is considered a rare delicacy. Harvesting this also kills the tree. Hearts of palm are often eaten in salads, sometimes called "millionaire's salad".
- Newly germinated coconuts contain an edible fluff of marshmallow-like consistency called coconut sprout, produced as the endosperm nourishes the developing embryo.

Non-culinary

- Coconut water can be used as an intravenous fluid.
- The water is also used in isotonic sports drinks.
- Coir (the fibre from the husk of the coconut) is used in ropes, mats, brushes, caulking boats and as stuffing fibre; it is also used extensively in horticulture for making potting compost.
- Copra is the dried meat of the seed and is the main source of coconut oil.
- The leaves provide materials for baskets and roofing thatch.

- Palmwood comes from the trunk and is increasingly being used as an ecologically-sound substitute for endangered hardwoods. It has several applications, particularly in furniture and specialized construction (notably in Manila's Coconut Palace).
- Hawaiians hollowed the trunk to form drums, containers, or even small canoes.
- The husk and shells can be used for fuel and are a good source of charcoal.
- Dried half coconut shells with husks are used to buff floors. In the Philippines, it is known as *bunot.*
- Shirt buttons can be carved out of dried coconut shell. Coconut buttons are often used for Hawaiian Aloha shirts.
- The stiff leaflet midribs can be used to make cooking skewers, kindling arrows, or are bound into bundles, brooms and brushes.
- The roots are used as a dye, a mouthwash, and a medicine for dysentery. A frayed-out piece of root can also be used as a toothbrush.
- Half coconut shells are used in theatre, banged together to create the sound effect of a horse's hoof beats. They were used in this way in the Monty Python film Monty Python and the Holy Grail.
- The leaves can be woven to create effective roofing materials, or reed mats.
- Half coconut shells may be deployed as an improvised bra, especially for comedic effect or theatrical purposes, such as in the 1970s UK sitcom It Ain't Half Hot Mum.
- In fairgrounds, a "coconut shy" is a popular target practice game, and coconuts are commonly given as prizes.
- A coconut can be hollowed out and used as a home for a rodent or small bird. Halved, drained coconuts can also be hung up as bird feeders, and after the flesh has gone, can be filled with fat in winter to attract tits.
- A 1.5" hole can be made in a coconut and a banana placed inside. Secured to a tree, it makes a monkey trap.

- Fresh inner coconut husk can be rubbed on the lens of snorkeling goggles to prevent fogging during use.
- Dried coconut leaves can be burned to ash, which can be harvested for lime.
- Dried half coconut shells are used as the bodies of musical instruments, including the Chinese yehu and banhu, and the Vietnamese ðàn gáo.
- Coconut is also commonly used as a herbal remedy in Pakistan to treat bites from rats.
- The "branches" (leaf petioles) are strong and flexible enough to make a switch. The use of coconut branches in corporal punishment was revived in the Gilbertese community on Choiseul in the Solomon islands in 2005.
- Coconut seedlings are popular novelty houseplants.
- In World War II, coast-watcher scout Biuki Gasa was the first of two from the Solomon Islands to reach the shipwrecked, wounded, and exhausted crew of Motor Torpedo Boat PT-109 commanded by future U.S. President John F. Kennedy. Gasa suggested, for lack of paper, delivering by dugout canoe a message inscribed on a husked coconut shell. This coconut was later kept on the president's desk, and is now in the John F. Kennedy Library.

Cultural Aspects

Coconuts are extensively used in Hindu religious rites. Coconuts are usually offered to the gods, and a coconut is smashed on the ground or on some object as part of an initiation or inauguration of building projects, facility, ship, etc.; this act signifies a sacrifice of ego, the idea that wealth stems from divinity, and the idea that, if due credit is not given, bad karma is taken on. In Hindu mythology it is referred as Kalpa vruksham. In Hindu mythologies it is said that *Kalapa vruksham* gives what is asked for.

- The Indonesian tale of Hainuwele tells a story of the introduction of coconuts to Seram.
- The people of the state of Kerala in southern India consider Kerala to be the "Land of Coconuts"; *nalikerathinte naadu* in the native language.

- The word "coconut" is also used as a mild derogatory slang word referring to a person of Latino, Filipino, or Indian subcontinent descent who emulates a white person (brown on the outside, white on the inside).
- "Coconut" is New Zealand slang for a Tongan, or other person of "Polynesian" descent, although usually not Maori.
- "Coconut" is also the title of a song by Harry Nilsson.
- "Coconut" is also the title of an In Reverie b-side track by Saves the Day.
- "Coconut" is also used as a slang term for breasts.
- Kid Creole's backing singers were known as his Coconuts.
- Cocolo originated as a term for a coconut seller.
- "Coconuts" is the title of a song by Widespread Panic.

Arecaceae

Arecaceae (sometimes known by the names Palmae or Palmaceae, although the latter name is taxonomically invalid.), the Palm Family, is a family of flowering plants belonging to the monocot order, Arecales. There are roughly 202 currently known genera with around 2600 species, most of which are restricted to tropical or subtropical climates. Most palms are distinguished by their large, compound, evergreen leaves arranged at the top of an unbranched stem. However, many palms are exceptions to this statement, and palms in fact exhibit an enormous diversity in physical characteristics. As well as being morphologically diverse, palms also inhabit nearly every type of habitat within their range, from rainforests to deserts.

Palms are one of the most well-known and extensively cultivated plant families. They have had an important role to humans throughout much of history. Many common products and foods are derived from palms, and palms are also widely used in landscaping for their exotic appearance making them one of the most economically important plants. In many historical cultures, palms were symbols for such ideas as victory, peace, and fertility. Today, palms remain a popular symbol for the tropics and vacations.

Bibliography

Adams H.: *Hot Pepper Improvement*, St. Michael, CARDI, 1997.

Agarwal P. K.: *Improvement of Citrus*, New Delhi, Malhotra Publishing House, 1993.

Alexander M. P. and Ganeshan S.: *Pollen Storage*, New Delhi, Malhotra Publishing House, 1993.

Andrews J.: *The Domesticated Capsicums*, University of Texas Press. 1995.

Andrews L.: *Citrus Production - Orange*, St. Augustine, Trinidad and Tobago, 1990

———: *Production of Minor Crops Sapodilla, Tamarind, Guava and Genip*, Port of Spain, Trinidad and Tobago, 1996.

Arguello D. S.: *Production, Post-harvest Management and Exportation of Tropical Fruits in Costa Rica*, Wageningen, Technical Centre for Agricultural and Rural Cooperation, 1992.

Ashworth S.: *Seed to Seed*, Decorah, Seed Savers Publications, 1991.

Barbeau G.: *Tropical Fruits in Nicaragua*, Managua, Nicaragua Ministerio de Desarrollo Agropecuario, Agraria, 1990.

Batlle I. and Tous J.: *Carob Tree (Ceratonia siliqua L.)*. Rome, International Plant Genetic Resources Institute, 1997.

Berger, P.L.: *Pyramids of Sacrifice: Political Ethics and Social Change*, New York, Basic Books, 1974.

———: *Pyramids of Sacrifice: Political Ethics and Social Change*, New York, Basic Books, 1974.

Brooks, D.: *Water: Local-Level Management*, Ottawa, International Development Research Centre, 2002.

Bunnik J. S. C.: *Fresh Fruits and Vegetable: a Survey on the Netherlands and other Major Markets in the European*

Community, Netherlands, Centre for the Promotion of Imports from Developing Countries, 1990.

Burton W.G.: *The Potato*, Holland, H. Veenman & Zonen N.V., 1966.

Cao Van P.: *An Integrated Approach for the Production and Processing of Minor Fruits in Martinique (FWI)*, Port of Spain, Trinidad and Tobago, 1996.

Cardona M. J. G., Carvajal S. L. B., Salinas D. G. C. and Isaza R. G. B. : *Proceedings of the International Seminar on Plantain Production, Quindio, Colombia,* Quindio, Corporation Colombiana de Investigation Agropecuaria, 1998.

Chadha K. L. and Pareek O. P.: *Advances in Horticulture: Fruit Crops,* New Delhi, Malhotra Publishing House, 1993.

Chambers, R.: *Rural Development: Putting the Last First*, London, Longman, 1983.

Collymore L.: *Fruit Production in Barbados*, Port of Spain, Trinidad and Tobago, 1996.

Coste R.: *Coffee: the Plant and the Product*, London, MacMillan, 1992.

Crucefix D.: *Avocado Variety Selection for Export Development,* Roseau, CARDI, 1996.

Cull Brian & Lindsay Pax: *Fruit Growing in Warm Climates for Commercial Growers & Home Gardeners*, Australia, Reed Books, 1995.

Currah L. and Proctor F. J.: *Onions in Tropical Regions*, Kent, Natural Resources Institute, 1990.

Daniells J.: *Illustrated Guide to the Identification of Banana varieties in the South Pacific*, Canberra, ACIAR, 1995.

Degras L.: *Yam: a Tropical Root Crop*, Wageningen, CTA/ MacMillan, 1993.

Dhatt A. S. and Singh Z.: *Propagation and Rootstocks of Citrus*, New Delhi, Malhotra Publishers, 1993.

Diederichsen A.: *Coriander (Coriandrum sativum L.).* Rome, International Plant Genetic Resources Institute, 1996.

Doijode S. D.: *Seed Germination in Fruits*, New Delhi, Malhotra Publishers, 1993.

Dudley, E.: *The Critical Villager: Beyond Community Participation*, London, Routledge, 1993.

Featherly H. I.: *Taxonomic Terminology of the Higher Plants*, USA, Iowa State College Press, 1954.

Ferentinos L.: *Proceeding of the Sustainable Taro Culture for the Pacific Conference*, Honolulu, HITAHR, 1993.

Forde S: *Proceedings of CARDI / CTA Workshop on Marketability of Caribbean Minor Fruits*, Port of Spain, Trinidad and Tobago, 1996.

Georges, S.: *The Debt Boomerang: How Third World Debt Harms Us All*, Boulder, Westview Press, 1992.

Glowinski Louis: *The Complete Book of Fruit Growing in Australia*, Australia, Lothian Publishing Company Pty. Ltd., 1991.

Godden G.: *Growing Citrus Trees*, Australia, Lothian Publishing Company Pvt. Ltd., 1988.

Gowen S.: *Banana and Plantains*, London, Chapman Hall, 1996.

Green S.K. and Kim J. S.: *Sources of Resistance to Viruses of Pepper (Capsicum spp.): a Catalogue*, Taipei, AVRDC, 1994.

Gunjate R. T. and Tawde A. B.: *Propagation and Rootstocks of Mango*, New Delhi, Malhotra, 1993.

Hartley W.: *A Checklist of Economic Plants in Australia*, Melbourne, C.S.I.R.O., 1979.

Harwood, R. R.: *Marshalling Technology for Development: Proceedings of a Symposium*, Washington, DC, National Academy Press, 1995.

Herklots G. A. C.: *Vegetables in South East Asia*, London, George Allen & Unwin Ltd., 1972.

Hessayon D. G. Dr.: *The Vegetable Expert*, England, PBI, Publications, 1985.

Huaman Z.: *Descriptors for Sweet Potato*, Rome, International Board for Plant Genetic Resources, 1991.

Hurst Jacqui & Rutherford Lyn.: *A Gourmet's Book of Mushrooms & Truffles*, Sydney, Golden Press Pvt. Ltd., 1991.

Jacquat Christiane: *Plants from the Markets of Thailand*, Bangkok, Duang Kamol, 1990.

Jeffers P.: *Evaluation of Four Onion Varieties in Montserrat*, Plymouth, CARDI, 1992.

Kenridge K. C. and Hardy B.: *Biology and Agronomy of Forage Arachis*, Cali, International Centre for Tropical Agriculture, 1994.

Kroll R.: *Cut Flowers*, Wageningen, CTA, 1995.

Kunelius T.: *Annual Ryegrasses in Atlantic Canada*, Ottawa, Agriculture Canada, 1991.

Lofgren, H., Richards, A.: *Food Security, Poverty, and Economic Policy in the Middle East and North Africa*, Washington, DC, International Food Policy Research Institute, 2003.

Mabberley D. J.: *The Plant-Book : a Portable Dictionary of the Vascular Plants*, Cambridge, Cambridge University Press, 1997.

Madulid Domingo A.: *A Pictorial Cyclopedia of Philippine Ornamental Plants*, Philippines, Makati Metro Manila, 1995.

Malins A.: *Postharvest Handling of Pineapple and Mango*, Port of Spain, Trinidad and Tobago, 1992.

Mannetje, L. T. & Jones, R. M.: *Plant Resources of South-East Asia,* Wageningen, Pudoc Scientific Publishers, 1992.

Matsuoka H.: *Cultivation of Panicum Genetic Resources for Evaluation of Characteristics*, Tokyo, JICA, 1997.

Mc Donald F.: *Plant Tissue Culture Manual for Yam, Cassava, Sweet Potato, Dasheen (taro) and Tannia (cocoyam)*, Roseau, Dominica and Cave Hill, 1993.

Mellor, J. W.: *The New Economics of Growth*, Ithaca, Cornell University Press, 1976.

Miller William: *Dictionary of English Plant Names*, London, John Murray, 1884.

Mitra S.: *Postharvest Physiology and Storage of Tropical and Subtropical Fruits*, Oxon, CABI, 1997.

Morris M. L. : *Maize Seed Industries in Developing Countries*, Colorado, Lynne Rienner, 1998.

Morton Julia F.: *Fruits of Warm Climates*, Miami, Julia F. Morton Publisher, 1987.

Murali T. P. and Duncan E. J.: *In Vitro Propagation of Banana through Aseptic Manipulation of Male Inflorescences*, Port of Spain, NIHERST, 1990.

Nabhan G. P.: *Wild Phaseolus Ecogeography in the Sierra Madre Occidental, Mexico: Areographic Techniques for Targeting and Conserving Species Diversity*, Rome, International Board for Plant Genetic Resources, 1990.

Nijdam J. & De Jong A.: *Elsevier's Dictionary of Horticulture in Nine Languages*, Elsevier Scientific Publishing Co. Amsterdam; New York.

Oka H. I.: *Origin of Cultivated Rice*, Elsevier, Japan Scientific Societies Press, 1988.

Oldham P.: *Cost of Production of Major Tree Crops in Dominica*, Roseau, Ministry of Agriculture, 1991.

Ou Yang Jue Ya: *Comparative Study of Mandarin and Cantonese*, Beijing, China Social Sciences Publishing House, 1993.

Percival John: *The Wheat Plant - A Monograph*, London, Duckworth & Co., 1921.

Pilgrim R.: *Post Harvest Handling of Minor Exotics*, St. George's, Grenada, 1996.

Ragone D.: *Breadfruit: Artocarpus Altilis (Parkinson) Fosberg*, Rome, International Plant Genetic Resources Institute, 1997.

Rehm Sigmund: *Multilingual Dictionary of Agronomic Plants*, Boston, Kluwer Academic Publishers, 1994.

Roy S. K.: *Research on Multipurpose Tree Species in Asia: In Vitro Clonal Propagation of Artocarpus Heterophyllus*, Bangkok, Winrock International Institute for Agricultural Development, 1991.

Singh H. P. and Chadha K. L.: *Genetic Resources of Citrus*, New Delhi, Malhotra Publishing House, 1993.

Sinha G. C., Reddy Y. T. N. and Singh G.: *Propagation and Rootstocks in Guava*, New Delhi, Malhotra Publishers, 1993.

Sperling L. and Berkowitz P. *Partners in selection: bean breeders and women bean experts in Rwanda*. Washington, USA: CGIAR, 1994.

Sreekumar V., Indrasenan G and Mammen G.: *Studies of the Quantitative and Qualitative Attributes of Ginger Cultivars*, Calicut, Kasaragod, 1990.

Stover R. H. and Simmonds N. W.: *Bananas*, United Kingdom: Longman Scientific and Technical, 1991.

Thavarasook C.: *Proceedings of the Mungbean Meeting 90, Chiang Mai, Thailand,* Bangkok, Tropical Agricultural Research Centre, 1991.

Thomas E.: *Fruit Production in St. Kitts and Nevis*, Port of Spain, IICA, 1996.

Vargas E. M., Macaya G., Baudoin J. P. and Rocha O. J.: *Variation in the Content of Phaseolin in Wild Populations of Lima Beans (Phaseolus lunatus L.) in the Central Valley of Costa Rica*. Plant Genetic Resources Newsletter, 2000.

Webb M.: *Weed Problems and their Control in Rice, Papaya, Citrus and Sugar cane: a Report to the Belize Plant Protection Service*, United Kingdom, Natural Resources Institute, 1993.

Whealy K.: *The Garden Seed Inventory*, Decorah, Seed Saver Publications, 1988.

Whitwell A.: *Dominica Orchard Crop Management and Research Project: Pest and Disease Management,* United Kingdom, Natural Resources Institute, 1991.

Woolfe Jennifer A.: *The Potato in the Human Diet*, Cambridge, Cambridge University Press, 1989.

Yoshida T.: *Cultivation of Citrus Genetic Resources for Evaluation of Characteristics*, Tokyo, JICA, 1996.

Index

A

B

C

H

I

L

M

N

O

P

R

S

T

U

V

W

❑❑❑